CONTENTS

HOW TO USE THIS BOOK

BEAST ACADEMY SCIENCE 3A

In your hands is the very first printing of the very first Beast Academy Science book ever written!

We are always creating new books, materials, and resources to complement this book.

Check the Science 3A page of BeastAcademy.com for the most up-to-date information and resources available.

PLANNING

Beast Academy Science 3A is the first in a two-part Level 3 science series. Together with 3B, Level 3 offers a complete science curriculum for motivated and curious 3rd graders.

Level 3 includes 32 chapters (16 per book). To complete the curriculum in a typical 36-week school year, we recommend completing 1 chapter each instructional week.

If time is limited, students may skip enrichment activities marked with a ✚ in the contents section at the beginning of each chapter.

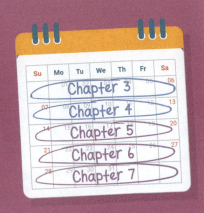

ACTIVITY TYPES

Comic Guide Sections AHA!

BA characters discuss important science concepts and ideas. Readers answer questions along the way.

Labs 🧪

One of the best ways to learn science is to do science! Students investigate how things work on their own.

Practice Pages

A variety of question types including puzzles, activities, and games help students learn to think like scientists.

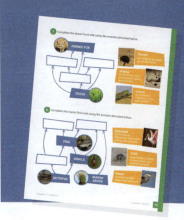

QUESTION TYPES

Every question students are expected to answer is numbered and has an answer or brief discussion in the back, beginning on page 318.

There are a wide variety of question types including ones marked by the icons below:

 **CHALLENGE:**
Starred questions and tasks have hints in the back starting on page 312.

 **DISCUSSION:**
Find a partner or group and work through these questions together.

 **RESEARCH:**
Students will need access to online resources to answer these questions.

 **JOURNAL:**
Collect answers to these questions in a separate science journal notebook.

STANDARDS

BA Science 3A content is aligned to the Next Generation Science Standards (NGSS), but often goes well beyond what is typically expected of students. Together with 3B, Level 3 will cover all of the concepts, practices, and standards outlined for grade 3.

Among the NGSS standards covered in this book are:

Forces and Interactions (Unit 1: Chapters 1-4)

Using patterns to predict motion (3-PS2-2)

Weather and Climate (Unit 2: Chapters 5-10)

Graphing seasonal weather data (3-ESS2-1)

Describing the world's regional climates (3-ESS2-2)

Reducing the impact of weather-related hazards (3-ESS3-1)

Relationships in Ecosystems (Unit 3: Chapters 11-16)

Exploring advantages of social behavior (3-LS2-1)

Recognizing how organisms adapt to survive in habitats (3-LS4-3)

Examining how changes to a habitat affect its organisms (3-LS4-4)

Inheritance and Variation of Traits (Unit 3: Chapters 11-16)

Identifying evidence that environment influences traits. (3-LS3-2)

Find additional information about the standards including core ideas, engineering practices, and crosscutting concepts online at ngss.nsta.org.

MATERIALS

SCIENCE JOURNAL

Every student should collect their work in a science journal. Any notebook will do.

COMMON ITEMS (Used in several activities)

- Scissors
- Tape
- Printer paper
- Stopwatch
- Thermometer
- Rubber bands
- Plastic water bottles
- Pens or Markers
- Cups

CHAPTER 1

- Marble
- Card stock or other heavy paper
- Rice or sand (optional)
- Variety of balls (optional)

CHAPTER 2

- Basketball
- Balloon
- Empty toilet paper rolls
- Feather or tissue
- Camera or phone with slow-motion video mode (optional)

CHAPTER 3

- Thin string
- Weights (such as metal washers)

CHAPTER 4

- Thin string
- Weights (such as metal washers)
- Measuring stick or measuring tape

CHAPTER 5

- Large baking sheet with rim
- Large bowl
- Toy car
- Towel (optional)

CHAPTER 7
- Orange
- Sharp pencil
- Headlamp (optional)

CHAPTER 8
- Flashlight
- Shoe box
- Dark construction paper
- Box cutter or knife (adult use only)

CHAPTER 9
- Air pump with inflating needle
- Nail similar in size to pump needle
- Hammer (adult use only)
- Marshmallows
- Bicycle tire
- Bicycle pump
- Push pin (optional)
- Glass bottle (optional)
- Ping pong ball (optional)
- Disposable coffee cup and lid (optional)
- Can of compressed air (optional)
- Camera or phone with slow-motion video mode (optional)

CHAPTER 10
- Two large containers
- Balloon
- Ice (optional)
- Thumb tack

CHAPTER 13
- Two tangerines or oranges

CHAPTER 15
- Coin
- Bubble wrap

CHAPTER 16
- Two sets of 6 tokens (we suggest 6 red and 6 black checkers)
- 30-36 dry beans or corn kernels
- Dice

Grogg (me)
"The Inventioneer!"

Invented the battery-powered hairbrush.

(And battery powered scissors to get it out of my hair.)

Winnie
"The firecracker"
Got mad when I told my see-food joke at lunch.

Lizzie
"The bookworm"
Reads a LOT! Even comics.

But she calls them "graphic novels."

Alex
"The executive"
Knows a LOT about spreadsheets.

Colored this drawing I made of him.

Mr. E.
Plays in a rock band!
Once fell asleep during a 15-minute xylophone solo. (More like Mr. Z's!)

Ms. Y.
Has lots of adventures.
Usually makes it back without getting lost.

Zak
Goes to the upper school.
Let me drive the remote-controlled sofa he built once. (Guess what - it floats!)

R&G
Take(s) care of the school.
I give the blue one fist bumps that explode.

Dr. Hau
Really old.
Like a bazillion years.
Says she always wanted to be an asteroid when she was a kid.

THE BEAST ACADEMY WORLD WORKS JUST LIKE OURS. THE CREATURES MAY LOOK DIFFERENT, BUT THE WAY THINGS WORK AND WHAT THEY'RE MADE OF ARE THE SAME AS HERE ON EARTH!

UNIT 1:
MOTION

The world around us is full of things that move, slide, fall, bounce, sway, tip, and turn in predictable ways. In this unit, we investigate objects in motion.

MOTION
CHAPTER 1:
Predictions

Have you ever seen an object that was about to fall and caught it just in time? How did you know where it was going to fall? You made a prediction without even thinking about it!

Watching objects in motion helps us see patterns that help us know how objects will move.

By the end of this chapter, you should be comfortable making predictions about objects in motion, and you should know more about why some things are harder to predict than others.

 =Lab    =Comic   ➕=Enrichment

CHAPTER 1: PREDICTIONS

Pitchers throw softballs at speeds over 60 miles per hour (100 km/h) that go right where they're aiming.

Scientists look for patterns in nature that help them make **predictions**.

The predictions we'll make in this chapter are about **objects in motion**.

Think about the last time you threw a ball. Did it land where you thought it would? If not, you can improve your aim with practice.

The better your aim, the easier it is to predict where the ball will go.

People have bowled more than 40 strikes in a row!

The world record for free throws made in a row is over 2,500!

CONTRAPTIONS LIKE THIS ARE SOMETIMES CALLED "RUBE GOLDBERG MACHINES" AFTER A CARTOONIST WHO DREW LOTS OF THEM.

TO **PREDICT** MEANS TO GUESS WHAT WILL HAPPEN BASED ON WHAT YOU'VE SEEN HAPPEN BEFORE. THE GUESS YOU MAKE IS CALLED A **PREDICTION**.

18

TARGET LAB

Like the little monsters in the Makerspace, you'll try to land a ball in a cup. How well can you predict where a marble will bounce?

MATERIALS

You will need:

- A marble
- Card stock, poster board, or other heavy paper
- Tape
- A cup
- A table on a hard floor
- A small stack of books
- Rice or sand (optional)

HITTING A TARGET

In this lab, our goal is to roll a marble down a ramp, off of a table, and bounce it into a cup.

If you do the same thing the same way each time, you should be able to **predict** where the ball will land.

Every time you roll the ball down the ramp, you learn a little bit more about where it will go.

But, it's not easy to get things exactly the same every time!

MAKE A RAMP

There are lots of ways to make a ramp. What is most important is that your marble rolls down your ramp the same way every time.

① Start with a rectangular strip of heavy paper.

② Fold the paper in half to make a V-shaped channel for your ball to roll in.

③ Fold the sides of the V down to make an M-shaped fold.

④ Fold a small piece of heavy paper in half. Tape one end of your ramp to the folded piece and tape the folded piece to the edge of a table as shown. Prop the top of your ramp on a small stack of books.

TIP:

Adding rice or sand in your cup will keep the cup from getting knocked over and will keep the marble from bouncing out of the cup.

ROLL YOUR MARBLE

Now, you're ready to roll!

Place a cup below the ramp where you think the marble will bounce. Place a marble at the top of the ramp.

Let it roll and see what happens.

OBSERVATIONS

What happened?

- The ball landed in the cup. Wow! You were either very careful or very lucky.
- The marble missed the cup completely.
- The marble bounced off of the side or rim of the cup.
- The marble bounced in and out of the cup.

Place the cup where you think the ball will land next time. When you place the cup, you're making a prediction about what you think will happen!

Roll the ball down your ramp 8 more times and record each roll in the table below. Before each roll, place the cup where you think it will land next.

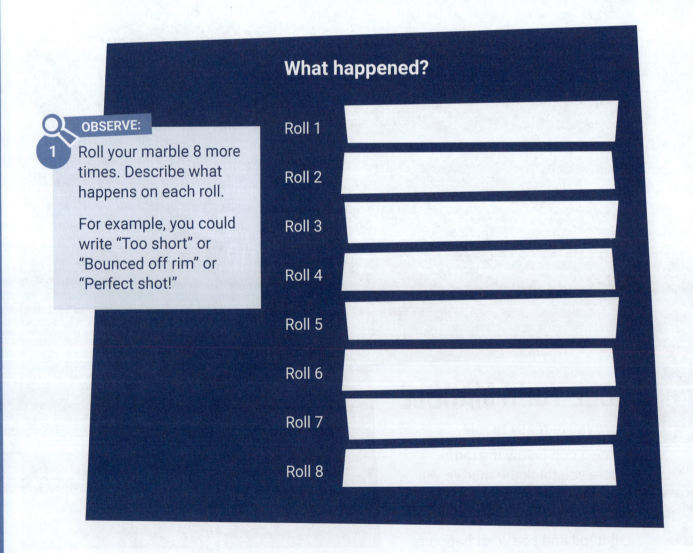

OBSERVE:

1 Roll your marble 8 more times. Describe what happens on each roll.

For example, you could write "Too short" or "Bounced off rim" or "Perfect shot!"

What happened?

Roll 1

Roll 2

Roll 3

Roll 4

Roll 5

Roll 6

Roll 7

Roll 8

Answer the questions below after rolling your marble at least 8 times.

2 Your marble probably didn't bounce exactly the same way every time. Describe the differences you noticed.

_____

_____

_____

3 List two or more things that you think could cause the marble to bounce a little differently and miss the cup.

_____

_____

_____

4 List two or more things you could try that might help you bounce the marble into the cup three times in a row.

_____

_____

_____

5 Can you bounce the marble into the cup more than three times in a row? What's your record?

EXTENSIONS

Whether you were able to bounce your marble into the cup 2 times in a row or 20, you probably learned that things don't always go exactly like you expect. But some things about the marble were very predictable. For example, it always fell down!

Try the extensions below to learn more about predictable motion.

6 Instead of bouncing your ball into a cup, try landing the marble in a cup without a bounce. Is this easier or harder than bouncing it into the cup? Why?

_____

_____

_____

7 Try rolling your marble from different places on your ramp. Does where you place your marble affect where it lands? How?

_____

_____

_____

8 Try changing the height of the top of the ramp. Does the height of the ramp affect where the marble lands? How?

_____

_____

_____

9 Try using other balls. Are some balls easier to predict than others? Is a ping pong ball easier or harder to predict than a marble? What about a bouncy ball?

_____

_____

_____

🖉 **JOURNAL:**

10 ⭐ Plan your own chain reaction contraption in your science journal. Or, try to build one!

A prediction is more than just a guess. Predictions are based on our observations of the world. You don't need to be a famous scientist to make predictions about objects in motion. All of us are always watching and learning how things work.

Use what you learned by watching your marbles and other rolling objects to help you make predictions.

PRACTICE:

Use what you know from observing the world to help you make a prediction for each pair of marbles. Write a sentence to explain your choice.

11 Which marble will hit the water first: A, or B? Why?

12 Which marble will land farther from the end of the ramp: C, or D? Why?

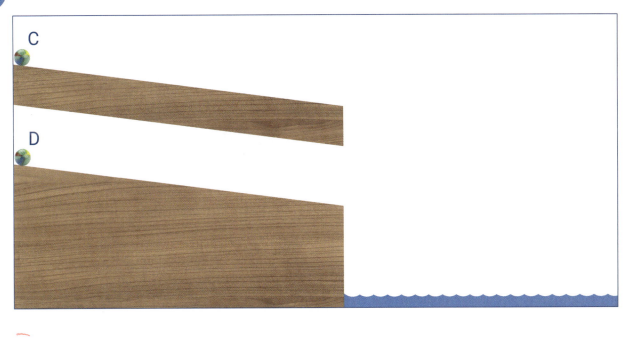

_____

_____

13 Which marble will hit the center block first: E, or F? Why?

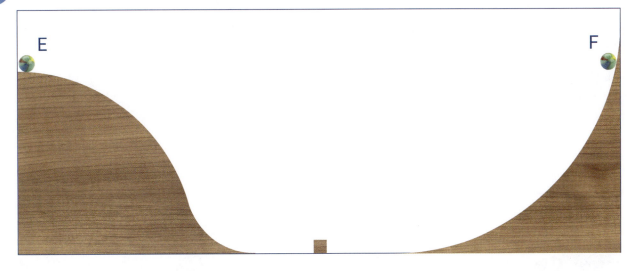

_____

_____

PREDICTING MOTION

In these problems, predict the path of each basketball. There may be more than one good answer.

PRACTICE:

Draw a line starting from the dot at the center of each basketball to describe the path you think the center of the ball will take after it is let go.

Example:

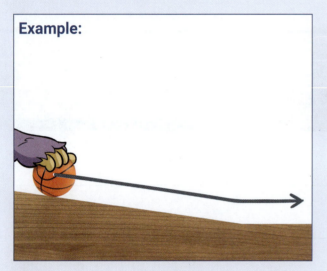

14

15

16

Some things are harder to predict than others
Remember, since you are drawing the path of the center of the basketball, your line should never touch the walls or the floor.

PREDICTABLE MOTION

Think about what makes some things easier to predict than others.

21 The three balls below are dropped from the same height onto a basketball court. Which ball's path do you think will be easiest to predict? Hardest? Explain.

Football   **Tennis ball**   **Bean bag**

_____

_____

_____

22 The three balls below are dropped from the same height onto the surfaces below. Which ball's path do you think will be easiest to predict? Hardest? Why?

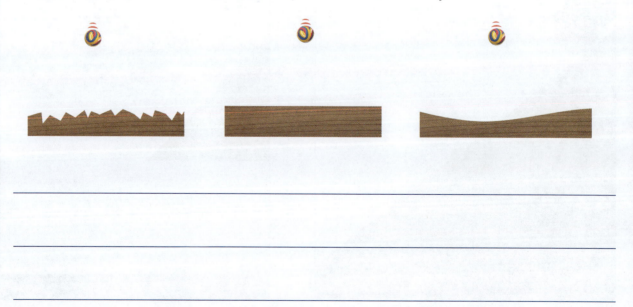

_____

_____

_____

SUMMARY

DID YOU KNOW?

The ancient Maya kept careful records of the Sun, the Moon, Venus, and Mars for hundreds of years. Their observations helped them predict events like eclipses far into the future without ever understanding how the Sun, Earth, Moon, and nearby planets move through space! They could predict eclipses without understanding what causes them!

Being able to make predictions is very important, but understanding why things happen is even better. Scientists always strive to understand why the predictions they make come true.

Watching what has happened helps us predict what will happen.

Some things are easier to predict than others. For example, it's much easier to predict what time the Sun will rise and set than to predict what time the next rain will start and end.

The more we observe the world around us, the more we can learn about the patterns and rules that nature follows.

A **solar eclipse** happens when the Moon passes between the Earth and the Sun. The Sun is about 400 times larger than the Moon, but also about 400 times farther away. So, the Sun and Moon look almost exactly the same size in the sky!

MOTION
CHAPTER 2:
Evidence

In the previous chapter, we made lots of predictions about objects in motion. How can we improve our predictions? It helps to know why things happen.

In this chapter, we'll gather evidence that will improve our predictions about falling objects.

By the end of this chapter, you should know why some objects fall through the air faster than others. You may even be able to predict how objects fall on the Moon!

🧪 =Lab   💥 =Comic   ➕ =Enrichment

CHAPTER 2:
EVIDENCE

In 1971, near the end of the Apollo 15 mission to the Moon, Commander David Scott dropped a hammer and a feather from the same height.

The simple experiment was recorded live and broadcast on television.

Which do you think landed first; the feather, or the hammer?

Why do you think this experiment was done on the Moon?

Investigating a prediction can help scientists gather **evidence** that helps them explain how the world works.

In this section, we'll gather evidence to investigate how objects fall.

CLAIMS & EVIDENCE

A **claim** is an explanation for something based on what we know. **Evidence** from observations and measurements can help us decide whether a claim is true, or help us revise it. Let's take a look at Grogg's claim about addition and multiplication below.

CLAIM:
Multiplying two numbers always gives a bigger result than adding them.

Grogg collects the evidence below to support his claim:

$5 \times 4 = 20$ $3 \times 3 = 9$ $7 \times 3 = 21$ $4 \times 9 = 36$ $6 \times 8 = 48$
$5 + 4 = 9$ $3 + 3 = 6$ $7 + 3 = 10$ $4 + 9 = 13$ $6 + 8 = 14$
$20 > 9$ $9 > 6$ $21 > 10$ $36 > 13$ $48 > 14$

1 Find two more numbers whose sum (+) is **less than** their product (×).

_____ _____

2 Find two numbers whose sum (+) is **greater than** their product (×).

_____ _____

3 Is Grogg's claim always true?

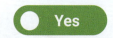 Yes   No

COUNTEREXAMPLES

One way to show that a claim is not always true is by finding a **counterexample**.

Grogg could write hundreds of examples that support his claim, but if he finds a counterexample, he will need to revise his claim.

PRACTICE:

Answer the questions below about counterexamples.

4   Find a counterexample that shows that the claim below is not always true.

> **All animals have teeth.**

5   What evidence would show that the claim below is not always true?

> **All rocks sink in water.**

When scientists discover evidence that shows their claim is not true, they can use what they've learned to change their claim.

DISCUSSION:

6   How could Grogg improve the claim he made on the previous page?

GATHERING EVIDENCE

Scientists design tests that will provide evidence for (or against) their claims.

To make a claim, it helps to have a question to answer. In this chapter, we'll explore claims and evidence to help us answer the following question:

DISCUSSION:

7 Below, Alex and Winnie have each made a claim that answers the question above.

How do the objects they are holding provide evidence that supports their claims?

PRACTICE:

Complete the tests below and answer the questions that follow.

Drop a large ball and a smaller balloon from the same height at the same time.

8 Which is wider: the ball or the balloon? _____

9 Which lands first: the ball or the balloon? _____

Cut or tear a sheet of paper into a small piece and a big piece. Crumple the small piece, but leave the big piece flat. Drop both from the same height at the same time.

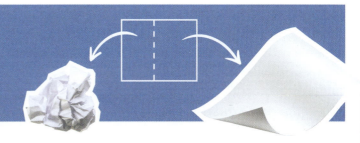

10 Which is heavier: the crumpled sheet or the flat sheet? _____

11 Which lands first: the crumpled sheet or the flat sheet? _____

JOURNAL:

12 Consider Winnie's and Alex's claims.

Which test above gives a counterexample to Alex's claim? Why?

Which test above gives a counterexample to Winnie's claim? Why?

Narrow objects fall faster than wide ones.

Heavy objects fall faster than light ones.

AIR RESISTANCE

Narrow objects usually fall faster than wide ones.

Alex and Winnie both had good ideas about what causes objects to fall faster or slower, but neither of their claims is completely true.

To understand why some objects fall faster than others, we need to understand how air affects falling objects.

For an object to fall, it has to push air out of the way. The push of air against a moving object is called **air resistance** or **drag**.

If you stick your hand out the window of a moving car, you can feel the air pushing against your hand. Turning your hand changes how much air your hand pushes. The less air you push, the easier it is to keep your hand where it is.

This helps explain why narrow objects usually fall through the air faster than wide ones.

Narrow objects don't have to push as much air out of the way when they fall.

PRACTICE:

Answer the questions below about air resistance.

13 Parachutes are used to slow the falls of people and objects. Why does a person fall so much slower with an open parachute?

_____

_____

14 Cyclists often race wheel-to-wheel in a line as shown. Why do you think cyclists ride like this?

_____

_____

DOES WEIGHT MATTER?

Think about how air affects objects of different weights that are about the same size and shape. A light breeze could blow a balloon for miles, but the same breeze wouldn't budge a bowling ball.

This helps explain why heavy objects usually fall through the air faster than light ones.

Heavy objects are not affected by air as much as light objects.

(This is due to a property called inertia. We explore inertia in Science 3B.)

> Heavy objects usually fall faster than light ones.

PRACTICE:

Circle the object you expect to fall faster, then circle the main reason why. Test your guesses by dropping objects like the ones below from the same height.

15 Which falls faster? (Mark one.)

Bag   **Crumpled Bag**

16 Why?

○ More weight

○ Less air resistance

17 Which falls faster? (Mark one.)

Bag   **Bag + Apple**

18 Why?

○ More weight

○ Less air resistance

19 Which falls faster? (Mark one.)

Apple   **Bag + Apple**

20 Why?

○ More weight

○ Less air resistance

SLOW YOUR ROLL

How can you slow an object's fall without changing its weight?

MATERIALS

You will need:

- Empty toilet paper rolls
  The more, the better
  At least one per person

- Scissors

- Camera or phone with a slow-motion video mode (optional)

This activity works great as a contest. Each participant begins with the same object. The goal is to modify it so that it falls as slowly as possible without adding or removing anything.

Empty toilet paper rolls are a good option, but you can use paper cups, large index cards, or any other common items.

Compete with yourself by making lots of different designs, or compete with a group to see who is best at slowing their roll.

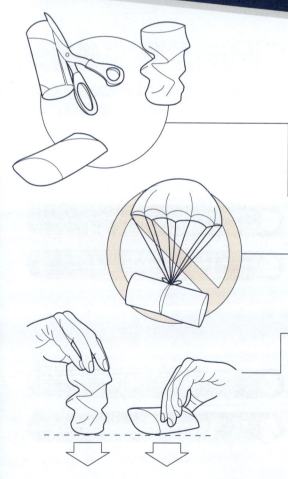

Rules

- You may fold, cut, bend, crinkle, crease, crimp, notch, twist, or tear your roll, but...

- You may not add or remove anything. Your roll must stay in one piece. No glue, no tape, no changing the weight.

- To compare two designs, the same person should hold both as high as possible with the bottom of each design at the same height. Release both at the same time.

- Last to hit the ground wins the round. Do at least 3 rounds to determine which is slower. If it's too close to tell, try using slow-motion video to help judge.

SEEDS FOR THOUGHT

We can look to nature for solutions to the problems we are trying to solve.

Many plants have seeds that fall slowly. The seeds can be carried great distances by the wind and spread out over a wide area.

Here are some examples of nature's slow-falling seeds that you can use as inspiration for your design.

Maple seeds have wings that cause the seeds to spin, slowing their fall.

The Javan cucumber plant has huge, 13 cm (5 inch) paper-thin seeds that fly like gliders as they fall.

These dandelion seeds dangle from tiny parachutes that allow them to fall slowly and travel for miles in the wind.

JOURNAL:

21 Include sketches of the best designs in your journal and explain why you think they worked well.

Ms.Y. Falling

Hey, little monsters.

Sorry I'm a little late!

Did you really go skydiving this morning!?

I was testing out a new flight suit design.

I've been gathering evidence on how it works.

What's it supposed to do?

It lets me change my shape to control how fast I'm falling.

Cool! How?

What do you think happens when I flatten myself out and extend these flaps?

What do you think?

BOTTLE DROP

We've seen evidence that both air resistance and weight can affect how fast an object falls.

In this demonstration, we drop two objects that are the same size and shape, but different weights. They land at the same time! Don't believe us? Try it!

MATERIALS

You will need:

- Two of the same plastic bottles
- Water
- Camera or phone with a slow-motion video mode (optional)

TIP:

Drop the bottles on a loud surface so you can hear each bottle land.

1. Fill one of the bottles to the top with water and fill the other only part of the way.

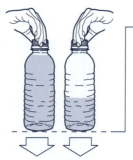

2. Hold both bottles at the same height outdoors. Drop both bottles at the same time. Watch and listen for which bottle lands first.

3. If possible, have a partner record the drop with a camera, using slow-motion mode if available.

4. Try this several times to gather evidence.

MORE TO EXPLORE

- Find a place where you can drop the bottles from higher, like a porch, a balcony, or a set of stairs. The higher, the better, but make sure the area below is clear.

- Compare bottles with different amounts of water. Try an empty bottle against a full one.

DISCUSSION:

22 Does this new evidence convince you that weight doesn't actually matter?

What's happening?

Unless you drop the bottles from pretty high up, both bottles probably hit the ground at the same time.

Unlike a feather or a balloon, both bottles are heavy enough that the air doesn't affect them much. For short drops, air doesn't slow either bottle down enough to notice.

Imagine both bottles on a table in the wind. A light breeze wouldn't do much to either bottle. It would take more wind to blow over the full bottle than the half-empty one. Air affects light objects more than heavy ones.

If you drop both bottles from 30 meters, they will begin to speed up. The faster they go, the more the air will push back—like a wind getting stronger and stronger. The air will slow down the half-empty bottle more than it will slow down the full one.

So, for longer drops, the difference between when the bottles land may be noticeable.

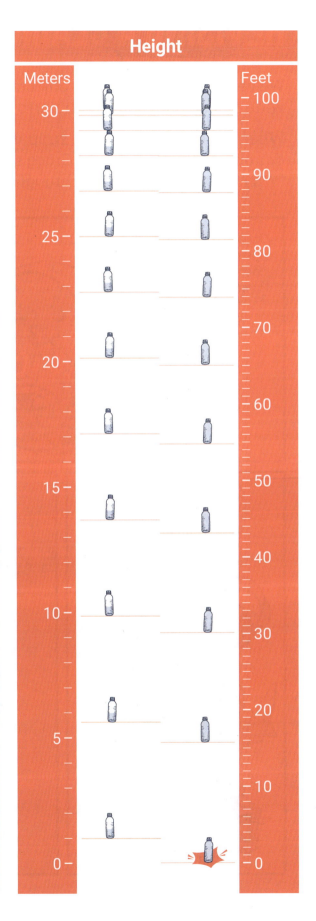

REVISING A CLAIM

We've seen that some light objects fall faster than heavier ones, and that some wide objects fall faster than narrow ones.

When the evidence doesn't support a claim, we can revise it—improve it based on our observations. Below is Grogg's revised claim to answer the question, "Why do some objects fall through the air faster than others?"

EVIDENCE FROM THE MOON

To test Grogg's claim that air causes objects to fall faster or slower, it would be helpful to see how objects fall when there is no air to slow them down.

This brings us back to the experiment on the Moon described at the start of this section.

What do you think happens to falling objects on the Moon, where there is no air to slow them down?

When Commander David Scott dropped a feather and a hammer from the same height in 1971, they landed at the same time!

You can watch the fuzzy video to see what the world saw more than 50 years ago. Search "Apollo 15 hammer feather drop" to watch the NASA video.

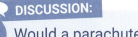
DISCUSSION:
23 Would a parachute help slow down a rocket landing on the moon?

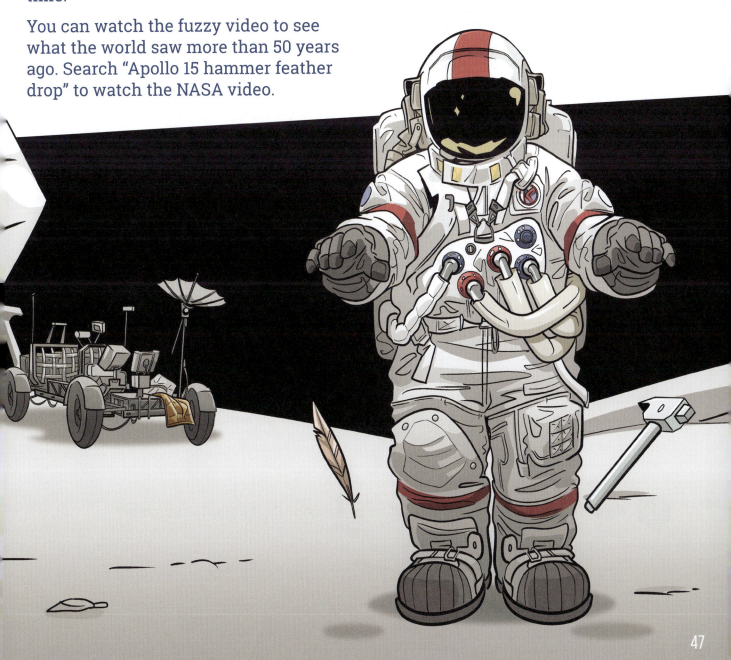

A FEATHER IN YOUR CUP

We can't remove all the air from the Earth, but there is a way to lower the air resistance on a feather. Then, we can see how fast a feather falls when air is not slowing it down.

MATERIALS

You will need:

- Disposable cup
- A feather or tissue

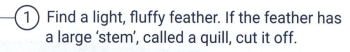

(1) Find a light, fluffy feather. If the feather has a large 'stem', called a quill, cut it off.

If you don't have a feather, cut the shape of a feather from a tissue.

(2) Hold the feather in one hand and a small cup in the other. Drop them both from the same height at the same time and watch how each falls.

(3) Next, hold the feather inside the cup as shown. Drop the cup and the feather at the same time and watch how each falls.

24 Which falls faster in step 2: the feather or the cup? Or, do they fall together?

_____

25 Which falls faster in step 3: the feather or the cup? Or, do they fall together?

_____

Answer the follow-up questions below.

26 When does the feather fall faster: in step 2 or step 3?

_____

27 Did the size, shape, or weight of the feather change between steps 2 and 3? If so, how?

_____

28 Does the air resistance on the feather change between steps 2 and 3? If so, how?

_____

29 Make a claim that answers the question, "Why does a feather fall faster or slower when it is in a cup?" Begin your claim, "The feather in the cup falls _____ because..."

_____

_____

JOURNAL:

30 **Follow-Up:**

Place the feather on top of a book as shown. Drop the book from a few feet on to a table, being careful to keep the book flat.

Make a prediction based on your claim above. Do the results support your claim?

MOTION
CHAPTER 3:
Cause and Effect

Have you ever wondered whether two events that seem unrelated are connected? Maybe you've seen someone sneeze right before a loud clap of thunder.

Did the sneeze cause the thunder? Probably not, but we could gather evidence to know for sure.

In this chapter, we'll investigate cause and effect relationships related to pendulums.

By the end of this chapter, you should know what changes you can make to a pendulum that will affect its swing.

 🧪 = Lab   AHA! = Comic   ➕ = Enrichment

CHAPTER 3:
CAUSE AND EFFECT

Scientists are always looking for **cause-and-effect** relationships that help them learn how one thing leads to another.

In the previous section, we investigated Alex's and Winnie's claims about what causes some objects to fall faster than others.

In this section, we will investigate why some things swing faster than others and make our own claims.

Galileo was one of the most famous scientists who ever lived.

He was especially interested in falling objects. Galileo made the claim that objects fall at the same speed without air resistance—a controversial discovery at the time that inspired the hammer and feather drop on the Moon over 300 years after his death (see page 47).

Galileo was also fascinated by swinging objects.

According to legend, when Galileo was a young man in Italy, he noticed patterns in the swinging lamps that were hanging from the ceiling. He wanted to understand what made the lamps swing the way they did.

Hanging lamps and chandeliers are types of **pendulums**—weights that can swing freely.

In this chapter, you'll investigate how pendulums swing, just like Galileo did.

A **COINCIDENCE** IS WHEN TWO UNRELATED EVENTS HAPPEN AT THE SAME TIME, USUALLY IN A SURPRISING WAY.

CAUSE AND EFFECT

An **effect** is any event or change that we can observe. A **cause** is what makes the effect happen.

Some cause-and-effect relationships are easy to discover.

Others are not as obvious.

Scientists look for cause-and-effect relationships that help them make predictions and even explain the reasons behind these predictions.

PENDULUMS

In this section we'll investigate cause-and-effect relationships for pendulums.

A **pendulum** is any hanging weight that can swing. Not all pendulums swing the same way. Some take longer to go back and forth than others.

What causes these differences?

What changes to a pendulum affect its swing time?

You've probably seen enough pendulums swing to make a few predictions before we start our investigation.

1. Which pendulum below do you think takes the most time to swing back and forth?

A   B   C

2. Describe at least two things about the pendulum you chose that might cause it to take longer to swing than the others.

_____

_____

MAKE A PENDULUM

To explore what causes a pendulum to swing the way it does, we'll make one of our own!

MATERIALS

You will need:

- String (The thinner, the better. Floss or thread works great.)

- Weights (We'll use metal washers.)

- Tape and scissors

- A stopwatch or other timer (Learn how to use it on pages 60-61.)

tape

MAKE A PENDULUM

To make a simple pendulum:

1. Cut a piece of string about the length of your arm.

2. Tie one end of the string to a large metal washer or other weight.

3. Tape the other end of the string below the edge of a table or anywhere else where there is room for it to swing without bumping anything.

SWING YOUR PENDULUM

Pull the weight at the end of your pendulum back and let it go (don't push it). It should swing back and forth smoothly without hitting anything.

1 FULL SWING

A full swing goes out and back.

The time it takes for a pendulum to complete one full swing is called its **period**.

Measure the time it takes for your pendulum to make ten full swings.

① Release the pendulum weight and start the stopwatch at the same time. Learn how to use a stopwatch on pages 60-61.

② Count 10 full swings. At the end of the tenth swing, stop the watch and record the time.

③ Careful! Don't say "one" until after the first swing is finished. It may help to say "zero" when you release the pendulum, then count to ten.

RESULTS:

3 How long does it take for your pendulum to make 10 full swings?

_____ seconds

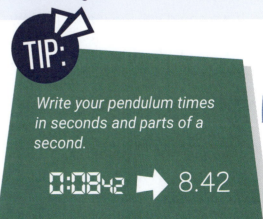

TIP:

Write your pendulum times in seconds and parts of a second.

0:08.42 ➡ 8.42

SCIENCE SKILL:
KEEPING TIME

You'll need to be able to measure time for some of the labs in this book. Practice using a stopwatch or other timer.

READING A STOPWATCH

The time on the stopwatch to the right shows 8 seconds and 42 parts of a second. We say, "8 point four-two seconds." We usually write a dot called a **decimal point** between the seconds and the parts of a second, like this:

8.42 seconds

The parts of a second are called hundredths because they split each second into one hundred tiny parts. The hundredths count from 00 to 99.

If we could see the watch in slow motion, it would say 8.43 seconds, then 8.44, then 8.45, and so on up to 8.99. Next is 9.00 seconds, then 9.01, and the hundredths count up to 99 again.

minutes   seconds   parts of a second

WRITING TIMES

We'll write times in minutes, seconds, and sometimes parts of a second. Write a colon between the minutes and seconds, and a decimal point between the seconds and the parts of a second.

Below are some examples of how we write and say the times we see on a stopwatch.

0:08ч2

8.42
Eight point
four two
seconds

2:0958

2:09.58
Two minutes,
nine point
five eight
seconds

0:32ч5

32.45
Thirty-two
point four five
seconds.

1: 10ɵ3

1:10.03
1 minute,
ten point
oh three
seconds

PRACTICE:

4 Practice using your stopwatch. Measure the time it takes for you to say the whole alphabet.

Do this five times. Try to say the alphabet about the same way each time.

Write the times, including the hundredths, in the blanks on the right.

RESULTS:
Timing the alphabet

1st Try: _____ seconds

2nd Try: _____ seconds

3rd Try: _____ seconds

4th Try: _____ seconds

5th Try: _____ seconds

INVESTIGATION

We will investigate cause-and-effect relationships with our pendulum.
Here are some important parts of any good investigation.

1

ASK A QUESTION

The first step in any investigation is figuring out what question you are trying to answer.

2

SET UP A TEST

Run a fair test that will help you answer your question. It's important to only change one thing at a time when looking for cause and effect relationships. If we change lots of things, we won't know which cause created the effect we observe.

3

GUESS

Making a guess helps you understand what you are looking for. Don't worry if your guess turns out to be wrong. We can learn a lot when an investigation doesn't give the results we expected!

4 PERFORM TRIALS

You'll want to run your test more than once. These tests are called **trials**. If you only test something once, it's hard to know if the same thing will happen again. The more we see the same thing happen, the more confident we can be in our results.

5 GATHER DATA

Collect data for each trial. If your data includes numbers and measurements, it helps to keep your data organized in a table like the one on the right.

Notes are data, too.

Bounce Height in Inches

	RED	GREEN
TRIAL 1	16	47
TRIAL 2	58	47
TRIAL 3	54	50
TRIAL 4	55	48
TRIAL 5	57	46

6 TAKE NOTES

Observations are more than just numbers and measurements. Include notes with your data, especially if there was anything weird or unexpected. That way, people who read the data will know what happened.

PENDULUM TEST 1: WEIGHT

A pendulum has two parts: a weight and a string.

First, we'll try changing the weight at the end of the pendulum to see if it changes the pendulum's swing time.

SETUP

We need two pendulums.

- The weight at the end is the only thing we should change. Try to keep everything else the same.
- The pendulums should be the same length, but one should have 1 washer and the other should have 3 washers.

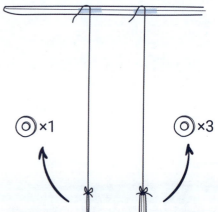

GUESS!

Guess what will happen before you collect any results.

5 Which do you think will take more time to make 10 full swings?

( ) 1 Washer   ( ) 3 Washers   ( ) About the same

6   Time both of the pendulums for 10 swings. Do this five times each.
Record your results and any notes below.

Time in Seconds (10 Full Swings)

	1 WASHER	3 WASHERS
TRIAL 1		
TRIAL 2		
TRIAL 3		
TRIAL 4		
TRIAL 5		

NOTES: _____

_____

_____

7   How do the times for the lighter weight compare to the times for the heavier weight?
Which swings took longer, or are they about the same?

○ **The lighter weight took longer**

○ **The heavier weight took longer**

○ **They were about the same**

DISCUSSION:

8   Which pendulum do you think would swing for longer before stopping? Test it!

PENDULUM TEST 2: STRING LENGTH

Next, we'll try changing the length of the pendulum's string to see if it changes the pendulum's swing time.

Does changing the length of a pendulum change the time it takes to make 10 swings?

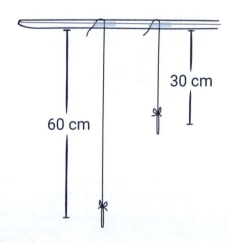

SETUP

We need two pendulums.

One should be about twice as long as the other when measured from the top of the string to the middle of the washer.

For example, one pendulum can be about 30 centimeters long (about 1 foot), and the other can be around 60 centimeters long (about 2 feet).

GUESS!

Guess what will happen before you collect any results.

9   Which pendulum do you think will take more time to make 10 swings?

 **Long String**   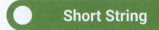 **Short String**    **About the same**

10 Time both of the pendulums for 10 swings. Do this five times each. Record your results and any notes below.

Time in Seconds (10 Full Swings)

	SHORT STRING	LONG STRING
TRIAL 1		
TRIAL 2		
TRIAL 3		
TRIAL 4		
TRIAL 5		

NOTES: _____

_____

_____

11 How do the times for the long string compare to the times for the short string? Which swings took longer, or are they about the same?

○ The longer string took longer

○ The shorter string took longer

○ They were about the same

DISCUSSION:

12 Do you think how far you pull the pendulum back affects its swing time? Test it!

CLAIMS

Now that we have investigated pendulums and gathered evidence about what changes affect the way a pendulum swings, we're ready to make some claims.

PRACTICE:

Choose the text that best completes the blank in each claim below.

13 Increasing the amount of weight at the end of a pendulum _____ .

- ( ) Increases the time it takes to swing
- ( ) Decreases the time it takes to swing
- ( ) Doesn't change its swing time much

14 Increasing the string length of a pendulum _____ .

- ( ) Increases the time it takes to swing
- ( ) Decreases the time it takes to swing
- ( ) Doesn't change its swing time much

15 Why is it important to only change one thing at a time when we are looking for cause and effect relationships?

_____

_____

DISCUSSION:

16 What else could you change about a pendulum that might affect the time it takes to swing back and forth?

Below are Grogg's results comparing pendulums that are the same except for the type of string they use. Answer the questions below about Grogg's results.

Time in Seconds (10 Full Swings)

	YARN	FISHING LINE
TRIAL 1	18.96	19.05
TRIAL 2	18.89	19.14
TRIAL 3	19.03	18.97
TRIAL 4	20.76	19.02
TRIAL 5	19.08	19.12

17 Grogg made a note that he lost count during one of his trials. Circle the trial above that you think Grogg lost count on.

18 Look at your data on page 65. How do you think the pendulums you used for your data were different from Grogg's? Explain.

_____

_____

DISCUSSION:

19 Based on Grogg's data, make a claim about whether the type of string makes a difference in a pendulum's swing time.

MOTION
CHAPTER 4:
Engineering

Engineers use science to make things that solve problems in the world.

In this chapter, you will use what you've learned in the motion unit to solve a problem that has been around since the beginning of time!

By the end of this chapter, you will be able to build a portable device that keeps time to the nearest second for at least one minute, and you'll understand more about how engineering connects science and invention.

🧪 = Lab   ➕ = Enrichment

CHAPTER 4: ENGINEERING

When scientists discover patterns and make preictions about what causes them, they can use what they know to solve problems.

Applying what you've learned to solve real-world problems is called **engineering**.

Pendulums have been used to solve lots of engineering problems.

The huge pendulum below helps keep one of the tallest skyscrapers in the world from swaying. At 508 meters tall (1,667 feet), Taipei 101 was once the tallest skyscraper in the world. It is built in an area where strong winds and earthquakes are common. So, it must be extremely stable. Engineers came up with a creative solution.

The pendulum below, which weighs 660,000 kilograms (almost 1,500,000 pounds), sways in a way that limits the movements of the building.

Engineering involves finding a problem to solve, brainstorming ideas, then building, testing, and tuning a design until you've created a solution you can present to others.

DEFINING THE PROBLEM

The first step in engineering is usually **defining a problem** to solve. We'll give you the problem this time, but always be on the lookout for problems that need solutions.

The problem we'll try to solve in this section is one that people have been working on for thousands of years — timekeeping!

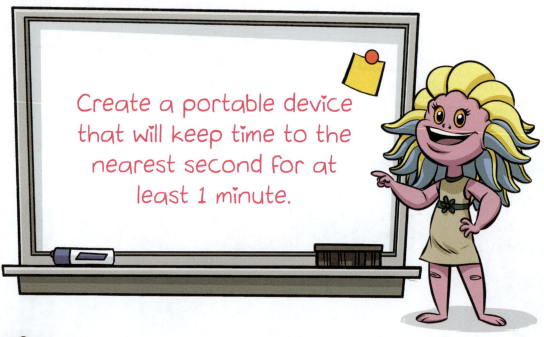

Create a portable device that will keep time to the nearest second for at least 1 minute.

BACKGROUND

For this engineering challenge, we'll be using pendulums to keep track of time.

The first pendulum clock was invented in 1656 by Dutch scientist and mathematician Christiaan Huygens. He was inspired by Galileo's work.

Pendulum clocks were the most precise way to keep time for more than 250 years, all the way until until the 1930's.

Pendulums are great for keeping time because their swings are very predictable.

RESEARCH

Engineers usually take time to **research the problem** next.

We spent the previous section researching pendulum swings. Let's review what we've learned that will help us engineer a timekeeping pendulum.

1 Describe at least one change you can make to a pendulum that significantly affects the amount of time it takes to make a full swing.

_____

_____

2 Describe at least one change you can make to a pendulum that does not significantly affect the amount of time it takes to make a full swing.

_____

_____

3 List any problems you encountered when building and testing a pendulum in the previous section that you will want to avoid this time.

_____

_____

RESEARCH:

4 Look at pendulums others have made to get ideas for how you'll make yours.

Search for "Science Pendulums" and "Pendulum Labs" to view images and videos of other pendulums and how to make them. You will probably find some cool ones!

BRAINSTORMING

Consider some things that will be important to the final design of your pendulum before you start building it.

PROBLEM 1: Portability

If it was the year 1656, you'd probably want to be able to take your pendulum everywhere to show off your new timekeeping device! Part of your engineering problem is making your timekeeper portable.

5 How can you make a pendulum that is easy to bring wherever you go?

What materials are available? How will you set up your pendulum? What will the top of your string attach to? How will it be attached? Sketch and write ideas for ways that you will be able to make your pendulum portable.

PROBLEM 2: Adjustability

When engineering a solution to a problem, it's good to be able to make adjustments easily.

6 How can you make a pendulum so that it is easy to adjust its length? Sketch or write ideas for ways that you will be able to make your pendulum longer or shorter. Consider ways to make both large and small adjustments.

Plan to make the length of your pendulum adjustable from 0 to 4 feet (0-122 cm).

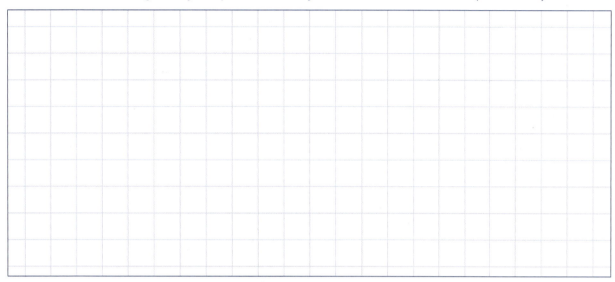

7 How can you make a pendulum so that it is easy to change the weight at the end? Sketch or write your ideas below.

Changing the weight may not affect the pendulum's swing time, but it might affect the pendulum in other ways that are important.

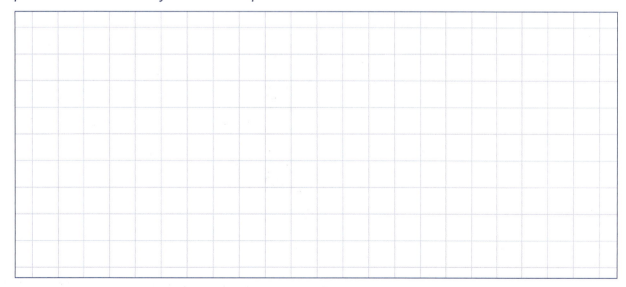

BUILDING

Now, it's time to put your best ideas together and build the first working model of your design, called a **prototype**. As you build, you may find ways to improve your design. Include these in your journal sketches like the ones below.

✏️ JOURNAL:

8 Add sketches of your design to your science journal. Include notes pointing out important parts. If possible, take photographs of your first prototype and attach them to a journal page.

Flying Dragon Pendulum *by Lizzie*

Heavy books keep it in place

String wraps around both push pins for big changes in length, or just one pin for smaller changes

Push Pins

Scrap wood from my mom's workshop.

Thin string

Shelf or dresser

The weight is a dragon toy. I can hang other dragons in the loop of string at the end if this one doesn't work.

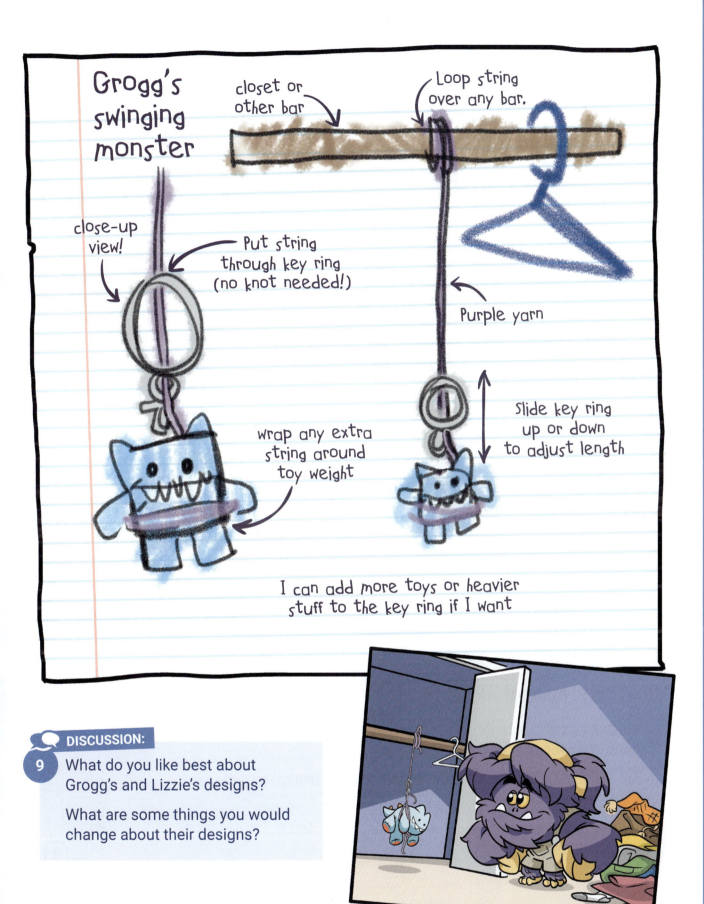

Grogg's swinging monster

closet or other bar

Loop string over any bar.

close-up view!

Put string through key ring (no knot needed!)

Purple yarn

wrap any extra string around toy weight

Slide key ring up or down to adjust length

I can add more toys or heavier stuff to the key ring if I want

DISCUSSION:

9 What do you like best about Grogg's and Lizzie's designs?

What are some things you would change about their designs?

TESTING YOUR PROTOTYPE

Build and test your **prototype** to make a pendulum that 'ticks' once every second.

MATERIALS

You will need:

- Your portable, adjustable pendulum

- A stopwatch or other timer (learn how to use it on pages 60-61)

- Measuring tape

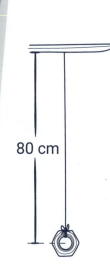

80 cm

SETUP

Start by setting up your pendulum and adjusting it to 80 cm.

When measuring the length of your pendulum, measure from the very top of the string to a point near the middle of the **pendulum weight**.

Measure to the same place on the pendulum weight every time.

COUNTING PENDULUM 'TICKS'

We will build a pendulum that swings out and back every 2 seconds (1 second in each direction).

We will call a swing in one direction a 'tick', like the tick of a clock.

The goal is to adjust your pendulum so that it completes 60 ticks in 60 seconds.

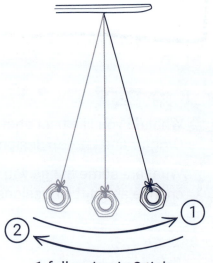

1 full swing is 2 ticks

Set up your pendulum and answer the questions below.

10 If a pendulum makes 60 ticks in 60 seconds, how long should it take to make 10 ticks?

- ○ 5 seconds
- ○ 10 seconds
- ○ 15 seconds
- ○ 20 seconds

11 Make sure your pendulum is 80 cm (about 32 inches) long and time it for 10 ticks. About how long do 10 ticks of your 80 cm pendulum take?

Your answer should be within 1 second of one of the values below. If not, try it again.

- ○ 6 seconds
- ○ 9 seconds
- ○ 12 seconds
- ○ 15 seconds

12 Will a pendulum that ticks once every second be longer or shorter than 80cm?

- ○ Longer
- ○ Shorter

13 Adjust your pendulum so that it is at least 10 cm longer or shorter and time it again. Continue to adjust your pendulum's length and time it for 4 more trials. As you get closer to the goal, make smaller adjustments to get close to one tick every second.

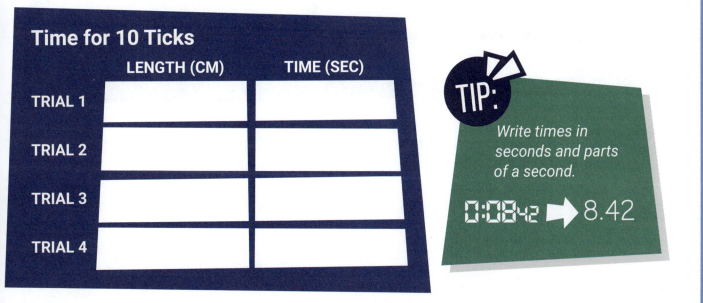

Time for 10 Ticks

	LENGTH (CM)	TIME (SEC)
TRIAL 1		
TRIAL 2		
TRIAL 3		
TRIAL 4		

TIP:

Write times in seconds and parts of a second.

0:08.42 ➡ 8.42

TUNING

By now, you probably have a pendulum that takes about 10 seconds to make 10 ticks. If not, make adjustments and complete a few more trials until your pendulum is ticking very close to once every second for 10 seconds.

Next, it's time to see how well your pendulum keeps time for a minute!

RESULTS:

Time your pendulum for 60 ticks, recording the length and time. Adjust your pendulum as needed until it ticks once every second, completing 60 ticks in 60 seconds.

14 Record at least 3 trials of 60 ticks of your pendulum in the table below. Make adjustments after each trial. We've given space for extra trials so you can get as close as you can to 60 seconds for 60 ticks.

Time for 60 Ticks

	LENGTH (CM)	TIME (SEC)
TRIAL 1		
TRIAL 2		
TRIAL 3		
TRIAL 4		
TRIAL 5		

15 Were you able to get your pendulum to tick once every second for 60 seconds? If not, what problems did you have? What changes could you make?

_____

_____

FIXING PROBLEMS

Your pendulum might not have worked exactly like you expected.

Below, we explore some problems that the little monsters had when testing their pendulum designs. Can you help them solve the problems they had?

PRACTICE:

Answer the questions below about problems that might come up.

16 Alex hangs his pendulum from a bar on the playground outside. His pendulum keeps blowing around in circles instead of swinging back and forth. What advice would you give to Alex about testing his pendulum?

_____

_____

17 No matter how far he pulls it back, Grogg's pendulum always stops swinging in less than a minute. What changes do you think Grogg could make to his pendulum that would make it swing for longer? (View Grogg's design on page 77.)

_____

_____

18 When Lizzie wraps her string around one of the push pins at the top of her pendulum (page 76), it changes the pendulum length by almost 2 cm. Why might this make it hard for Lizzie to adjust her pendulum to complete 60 ticks in exactly 60 seconds?

_____

_____

MORE TO TRY

Engineers never run out of things to build.

If you successfully created a pendulum that ticks once every second for at least 60 seconds, share it with others. Explain what it does and how it works.

Then, choose one or more of the challenges below to try.

 19 Make a pendulum that will keep time for much longer than one minute.

How long can you get a pendulum to tick once every second?

 20 Make a pendulum that ticks every 2 seconds (4 seconds out-and-back).

It will be taller than most ceilings are high, so find a tall space if you want to attempt this challenge. How does the length of a 2-second pendulum compare to the length of a 1-second pendulum?

 21 Make a double pendulum. Hang a heavy weight to make a standard pendulum. Then, hang a much lighter pendulum from the heavy weight to make a second pendulum of the same length.

Swing the lighter weight and watch what happens.

 22 Make a coupled pendulum. Hang two pendulums that are the same length from the same rope as shown on the right.

Swing one of the pendulums and watch what happens.

OTHER AMAZING PENDULUMS

There are dozens of cool pendulum projects you can try.

The projects below require help from a very patient and curious adult with tools. Most are best completed in a garage or other space where things can get messy.

Fill a container with sand, water, or paint. Poke a hole in it and and swing it in big arcs above a surface to create art.

Or, make a heavy pendulum with a pointed tip that traces similar patterns in sand as shown on the right.

Search "Sand Pendulum" or "Pendulum Painting" to find amazing **pendulum art**.

A **flying pendulum** is a device invented for use in clocks. Instead of simply swinging back and forth, the powered pendulum wraps itself around rods the way a tetherball wraps around a pole.

Search "Flying Pendulum" to find videos of these clever devices.

A **pendulum wave** is made with multiple pendulums of varying lengths that are carefully calculated so that each pendulum has a specific period. Swung together, the pendulums create mesmerizing patterns.

Search "Pendulum Wave" to find videos of these incredible pendulums.

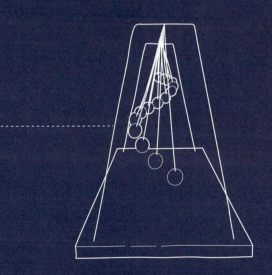

FOUCAULT PENDULUMS

If you have ever seen an enormous pendulum swinging from a high ceiling and knocking down small pegs throughout the day, you were probably looking at a Foucault (foo-coh) pendulum.

These giant pendulums are named after French physicist Léon Foucault. In 1851, Foucault suspended a 28-kilogram (62 lb) weight on a 67-meter (220 ft) wire from the dome of the Pantheon in Paris, France. He set it swinging to demonstrate the rotation of the Earth. We'll explain how on the next page.

Pictured below is a copy of Foucault's pendulum. It hangs exactly where the original was first set up in Paris.

Even huge pendulums don't swing forever. Some are restarted regularly. Others use a special device that gives the pendulum a little push without affecting the direction of its swing.

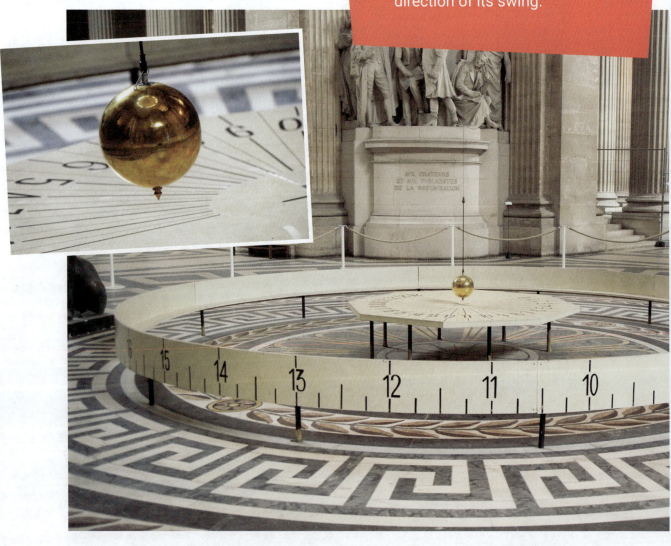

WHAT DOES IT DO?

A **Foucault pendulum** does what any other pendulum does—it swings back and forth. But, to a person watching the pendulum, it looks like the pendulum's swing is slowly rotating. The path of the weight changes during the day as shown on the right.

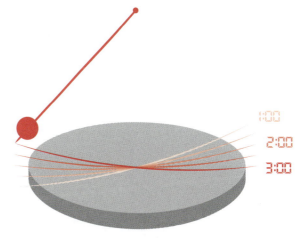

1:00
2:00
3:00

But, a person watching from space would see something different. Imagine a huge pendulum suspended at the North Pole. Watching from space, the pendulum swings left and right along the same path all day as the Earth turns below it.

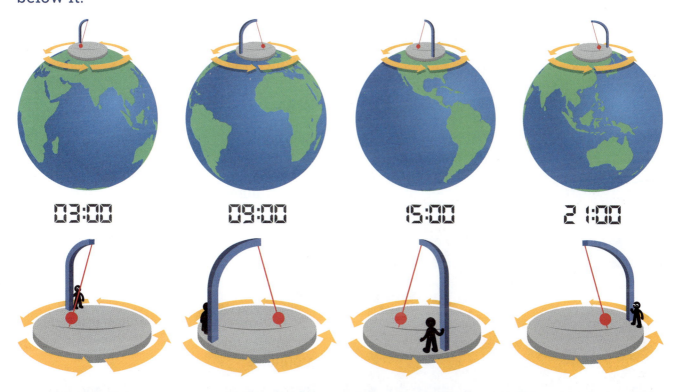

03:00   09:00   15:00   21:00

What's going on?

The pendulum's swing looks like it's rotating because the Earth is spinning beneath it!

The top of a Foucault pendulum is attached in a way that doesn't twist the string as the Earth spins. A big, heavy pendulum tends to keep swinging in whatever direction it started.

So, even as the floor, the building, you, and everything else around it rotates, the pendulum weight just keeps going back and forth the same way.

It's actually the Earth and everything on it that is rotating, not the pendulum!

UNIT 2:
WEATHER

Weather is how the air around us feels right now. In this chapter we explore patterns and differences in weather around the globe and learn some of the causes of these differences.

WEATHER
CHAPTER 5:
Temperature & Precipitation

Is 15°C a comfortable outdoor temperature?

Is 3 centimeters a lot of rain for one day?

The answers to these questions probably depend on where you live and what you're used to.

Temperature and precipitation are two common ways to describe the weather around us.

By the end of this chapter you should know how temperature and precipitation are measured and what these measurements mean.

🧪 =Lab   ➕ =Enrichment

CHAPTER 5:
TEMPERATURE & PRECIPITATION

Weather describes the current conditions in the air around us. Two important measures of weather are **temperature** and **precipitation**.

Death Valley is a desert valley in California that is one of the hottest and driest places on Earth. The temperature can get above 131°F (55°C), and it normally only rains about 60 mm each year (just over 2 inches).

How hot does it get where you live? How much rain and snow falls in a year? How do these numbers compare to the extremes of Death Valley, California?

TEMPERATURE

Temperature is the measure of how hot or cold something is.

We can describe air temperature with words like cold, cool, comfortable, warm, and hot. Or, we can measure temperature in degrees using a thermometer like the one below.

Most thermometers in the U.S. show temperatures using two types of degrees: degrees Celsius (°C) and degrees Fahrenheit (°F).

Air temperature is not the only thing that makes us feel hot or cold. Wind, shade, humidity (moisture in the air), and many other factors affect how a temperature feels.

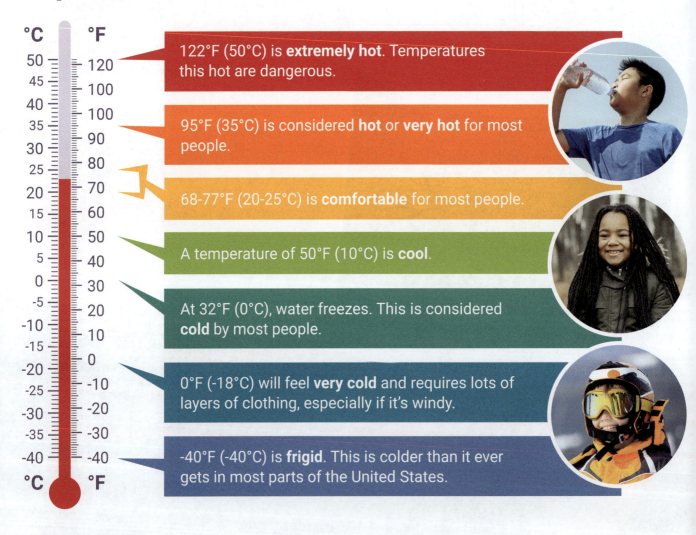

122°F (50°C) is **extremely hot**. Temperatures this hot are dangerous.

95°F (35°C) is considered **hot** or **very hot** for most people.

68-77°F (20-25°C) is **comfortable** for most people.

A temperature of 50°F (10°C) is **cool**.

At 32°F (0°C), water freezes. This is considered **cold** by most people.

0°F (-18°C) will feel **very cold** and requires lots of layers of clothing, especially if it's windy.

-40°F (-40°C) is **frigid**. This is colder than it ever gets in most parts of the United States.

 JOURNAL:

1 What is the outdoor temperature where you are right now in °F and °C? How does it feel?

PRACTICE:

Use the thermometer on the previous page to help you fill in the missing temperatures below. Then, write a short description like "warm," "hot," or "very cold".

2

15 °C ______ °F

Description: ______________________

3

______ °C 86 °F

Description: ______________________

4

-5 °C ______ °F

Description: ______________________

5

______ °C 41 °F

Description: ______________________

6

______ °C -4 °F

Description: ______________________

7

20 °C ______ °F

Description: ______________________

Solutions: 326-328

The table below shows the high temperature for a normal day in each month in three U.S. cities. Use the table to help you answer the questions below.

Normal Daily High Temperatures by Month (°F)

	Jan	Feb	Mar	Apr	May	Jun	Jul	Aug	Sept	Oct	Nov	Dec
Seattle	47°	49°	52°	57°	63°	66°	72°	73°	67°	59°	51°	46°
Chicago	32°	36°	45°	56°	66°	77°	82°	81°	74°	62°	50°	37°
Phoenix	68°	72°	77°	86°	94°	104°	106°	105°	100°	89°	76°	68°

8. What is a normal high temperature in August for each of the cities in the table?

Seattle ☐   Chicago ☐   Phoenix ☐

9. What is a normal high temperature in January for each of the cities in the table?

Seattle ☐   Chicago ☐   Phoenix ☐

10. If we call a month "comfortable" when the normal high temperature is between 60 and 80 degrees, how many comfortable months does each city above have?

Seattle ☐   Chicago ☐   Phoenix ☐

11 Look up the normal high and low temperatures for where you live and write them in the table below.

Search for "monthly temperatures" and your city or zip code.

Normal Daily High and Low Temperatures by Month (°F)

Location:

	Jan	Feb	Mar	Apr	May	Jun	Jul	Aug	Sep	Oct	Nov	Dec
High												
Low												

12 How do temperatures where you live compare to the cities in the table on page 90? Do you prefer the temperatures where you live, or in one of the cities in the table?

What is a "normal" day?

The charts and graphs in this chapter often show "normal" values for temperature and precipitation. When we say that the normal high temperature for Chicago in November is 50°F, that doesn't mean that every day in November will reach exactly 50°F. Some days will be hotter and some days will be colder.

A normal high of 50°F in November means that the high temperatures balance around 50°F. For example, it may reach 45°F one day and 55°F the next. 50°F is called the average high temperature.

To find the average daily high temperature for November in a city, you add up the high temperatures for every day

of the month, then split this amount equally among all the days. Since November might be hotter some years than others, it's good to do this over many years to get the best idea of an average or "normal" high temperature for a city in a particular month.

TEMPERATURES IN THE U.S.

Depending on where you live, 59°F (15°C) might be the temperature on a cool day in winter, or on a warm summer afternoon.

The map below shows normal daily high temperatures in April across the U.S. in degrees Fahrenheit. The tiny areas in light blue are the coldest, while the darkest orange areas are warmest. If you live in the U.S., find where you live on the map and think about what the weather there is like in April.

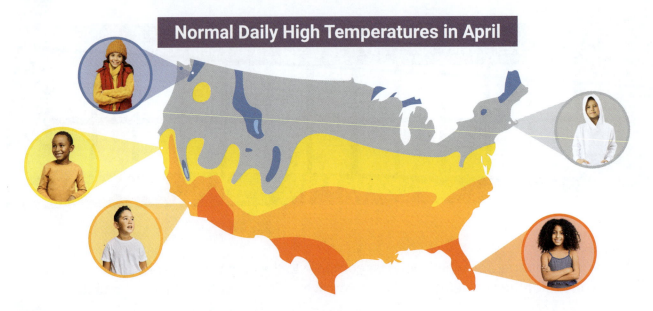

Normal Daily High Temperatures in April

13 The colors on the map stand for Fahrenheit temperatures in a ten-degree range. Label each color below with the temperatures you think it stands for. For example, 50's, 60's, or 70's.

14 How do you think this map would look different if it showed high temperatures in May instead of April?

_____

_____

Answer the questions below about temperature changes during the day where you live.

15 What times of day do you think are *usually* coldest? What times are hottest? Why?

_____

_____

RESEARCH:

16 Look up the hourly forecast for where you live. Fill in the temperatures for tomorrow's forecast, starting at midnight, in the table below.

Search for "Hourly Forecast" and your city or zip code.

	12:00 am	2:00 am	4:00 am	6:00 am	8:00 am	10:00 am	12:00 noon	2:00 pm	4:00 pm	6:00 pm	8:00 pm	10:00 pm
Temp.												

17 When are temperatures lowest in your chart above? When are temperatures highest?

Lowest _____   Highest _____

JOURNAL:

18 Use a thermometer to measure the actual temperature in the shade for at least three of the times given in the forecast above.

How closely do the forecast predictions match your measurements?

PRECIPITATION

Precipitation is liquid or frozen water that falls from the sky to the ground.

Most precipitation falls as rain. Rainfall is measured as the depth of water that reaches the ground, usually in inches (in) or millimeters (mm). For example, if it rains 3 inches during a storm, all of the containers below will fill to a depth of 3 inches.

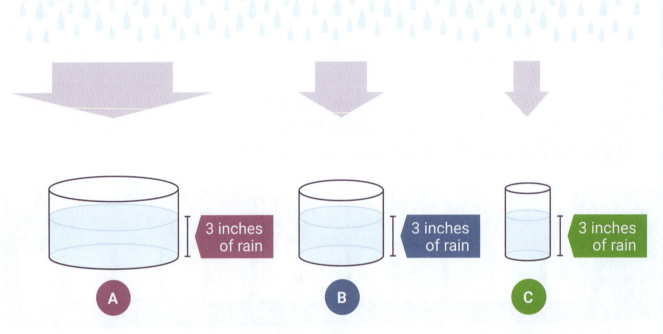

Imagine covering a whole parking lot with containers like these during a storm. Every container would fill to about the same depth!

19 All three containers above collect rain from the same storm. So, they all fill to the same depth. But, they don't hold the same amount of water. Which holds the most? Why?

_____

_____

_____

20 Alex leaves an empty cup next to his swimming pool during a rain storm. After the storm, the cup has 2 inches of water in it. How much did the water level in Alex's swimming pool rise?

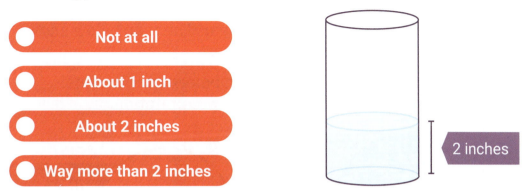

- ⬤ Not at all
- ⬤ About 1 inch
- ⬤ About 2 inches
- ⬤ Way more than 2 inches

2 inches

21 Did more water fall in Alex's cup or in Alex's swimming pool? Explain.

_____

_____

22 The containers below were outside Winnie's and Lizzie's houses during a storm. Whose house got more rain? Explain.

Winnie

Lizzie

_____

_____

✏️ **JOURNAL:**

23 Leave a cup out on a day when it rains. Measure the depth of the rain that collects in the cup. Did the cup collect more or less rain than you expected?

PRECIPITATION IN THE U.S.

Depending on where you live, 5 inches of rain might fall in a single weekend storm, or it might only rain 5 inches in a whole year! The map below shows the normal annual rainfall across the United States.

Snowfall is included as precipitation in the chart below after it melts. 10 inches of snow is usually about 1 inch of liquid precipitation.

Seattle, Washington
Rainy days: 150
Yearly rainfall:
40 inches (102 cm)

Chicago, Illinois
Rainy days: 125
Yearly rainfall:
35 inches (89 cm)

Phoenix, Arizona
Rainy days: 30
Yearly rainfall:
9 inches (23 cm)

New Orleans, Louisiana
Rainy days: 50
Yearly rainfall:
65 inches (165 cm)

Less than 5 inches (13 cm) per year.
5-12 inches (13-30 cm) per year.
12-20 inches (30-51 cm) per year.
20-30 inches (51-76 cm) per year.
30-40 inches (76-102 cm) per year.
40-50 inches (102-127 cm) per year.
50-70 inches (127-178 cm) per year.
Over 70 inches (178 cm) per year.

Mt. Waialeale, Hawaii
The rainiest place in the U.S. gets about 450 inches (over 11 meters) of rain each year. It is one of the rainiest places on Earth!

Answer the questions below about precipitation.

24 Look up the annual rainfall in a city near you. How many inches of water fall in a normal year where you live?

Search for "annual rainfall" and the nearest city.

inches

25 How many days does it rain or snow in a normal year where you live?

Search for "annual rainy days" and the nearest city.

days

26 Look up the normal monthly precipitation amounts for where you live and write them in the table below.

Search for "monthly precipitation" and your city or zip code.

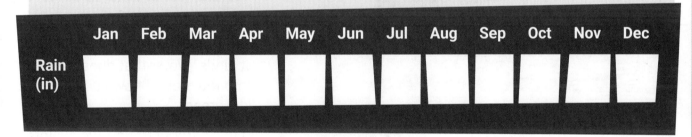

	Jan	Feb	Mar	Apr	May	Jun	Jul	Aug	Sep	Oct	Nov	Dec
Rain (in)												

27 Compare Seattle and New Orleans on the previous page. Which city gets more inches of rain each year? Which city has more rainy days?

_____

_____

28 What do you think is different about the rainy days in Seattle and New Orleans? How do these cities compare to where you live?

FLOOD DEMO

How can just a few inches of rain cause a flood?

MATERIALS

You will need:

- Water
- Small cup
- Large, rimmed baking sheet
- Large bowl
- Toy car
- Towel (for the extension)

FLOODS

A few inches of rain can cause several feet of water to collect in some areas. How?

Water doesn't usually stay where it lands. Water that doesn't soak into the ground flows downhill into low areas.

In this demonstration, we explore how a small amount of rain can cause a flood that could wash away cars!

① Pour two small cups of water onto your baking sheet. Place the toy car in the water on the baking sheet.

✏ **JOURNAL:**

29 Make a drawing that shows how the depth of the water on the baking sheet compares to the height of the toy car. Is the water over the tires? The hood? The roof?

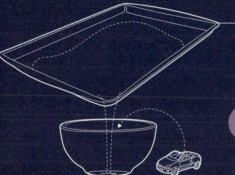

② Next, with help from an adult, carefully pour the water from one corner of the baking sheet into the big bowl. Place the toy car in the bowl.

✏ **JOURNAL:**

30 Make a second drawing that shows how the depth of the water in the bowl compares to the height of the toy car. Is the water over the tires? The hood? The roof?

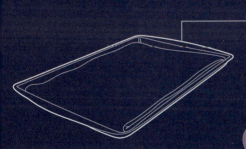

(EXTENSION) Start over. Try the two steps above, but this time place a folded towel on the baking sheet before you pour any water on it. Pour the water onto the towel-covered baking sheet, then into the bowl.

✏ **JOURNAL:**

31 What happens when you pour the water from the towel-covered baking sheet into the bowl? How do you think this relates to flooding in the real world?

Solutions: 326-328

FLOOD PREVENTION

In the flood demonstration on the previous pages, we were able to keep the water from flowing out of the pan and into the bowl by covering the pan with a towel.

The towel soaked up most or all of the water and held it in place so it couldn't flow quickly into the bowl. We can't cover cities in towels to prevent flooding, but the towel can teach us about one method of flood prevention.

PRACTICE:

Answer the questions below about rainfall and flooding.

32 Circle the surface below that you think will soak up the most rain water. Cross out the surface that you think will soak up the least water.

Brick Pavers

Tall Grass

Loose Soil

Paving Blocks

33 Circle the surface below where rain water will flow slowest down a hill. Cross out the surface where rain water will flow fastest.

Dense Plants

Grass Lawn

Concrete Sidewalk

Packed Dirt

34 How can having natural areas with lots of plants help reduce flooding in a city?

_____

_____

DISCUSSION:

35 Which surfaces on the previous page could be replaced with a different surface to reduce flooding? Which replacements wouldn't make sense?

JOURNAL:

36 This building in Rome, Italy has a roof covered with plants and soil. It's called a green roof.

How could buildings with green roofs help reduce flooding in a city?

What are some other possible benefits and problems of a roof like this?

Sponge Cities

A **sponge city** has lots of natural areas that absorb rain and help prevent flooding.

Collecting storm water is just one of the many benefits of having lots of natural areas in a city.

COLLECTING RAIN

Rainfall is measured using a **rain gauge** which collects water in a tube like one of the containers shown on the right.

Both containers have the same-sized opening in the top. So, they will collect the same **total amount** of rain.

But the water **depth** will be higher in the container with the smaller collection tube.

Most rain gauges are shaped like the container on the right. Since the tube on the right fills up faster, it's easier to tell the difference between small amounts of rain.

PRACTICE:

Think about how the size of each container compares to the opening in the top as you answer the questions that follow.

---

 37 The three containers below collect rain in the same location. Which one do you think will fill up first? Which one do you think will fill up last? Why?

_____

_____

_____

It may help to discuss these with a partner.

38 The three containers below collect rain in the same location. Which one do you think will fill up first? Which one do you think will fill up last? Why?

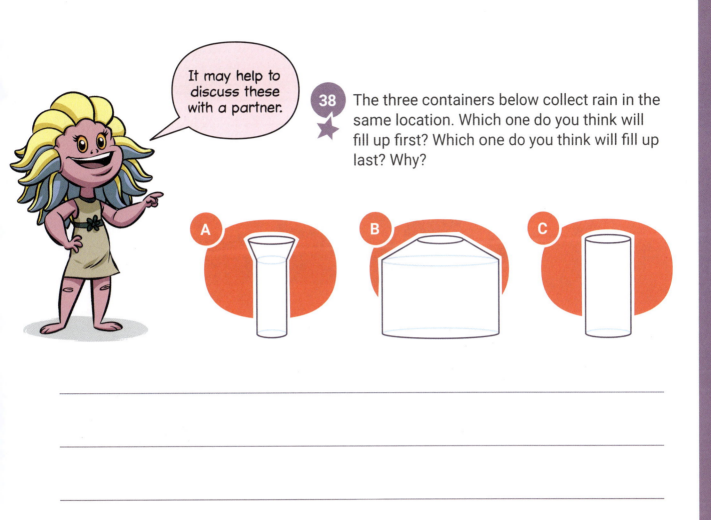

A   B   C

_____

_____

_____

39 The three containers below collect rain in the same location. Which one do you think will fill up first? Which one do you think will fill up last? Why?

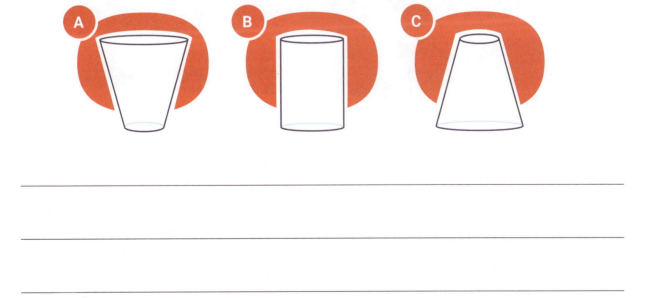

A   B   C

_____

_____

_____

FROZEN PRECIPITATION

Precipitation isn't always liquid. Frozen water can fall as snow, hail, or sleet.

SNOW

Snow is made up of frozen ice crystals called snowflakes that form in clouds and fall to the ground.

Just 1 centimeter of rain can fall as 10 or more centimeters of snow.

Some parts of the world like this road in Japan can receive over 12 meters (about 40 feet) of snow in a single winter!

40 The bucket on the right is left out in a snow storm and fills completely with snow. After all of the snow in the bucket melts, about how high will the water fill the bucket? (Circle one.)

A
B
C
D

HAIL

Hail stones are balls of ice that form in storms. Wind carries water drops upward into very cold air where they collect and freeze together. The updrafts can keep these balls of ice growing in the air until they are too heavy and fall to the ground. Hail can be anywhere from 5 millimeters (●) to 15 centimeters (about 6 inches)!

SLEET

In the U.S., when raindrops fall through cold air and freeze into ice pellets, these pellets are called sleet.

In other places, like Canada, frozen raindrops are called ice pellets. There, the word sleet is used to describe a mix of rain and snow.

Sleet pellets are smaller than 5 millimeters (●) across.

41 Circle every object below that could be the same size as a hail stone.

Salt Grain   **Glass Marble**   **Golf Ball**   **Tennis Ball**   **Soccer Ball**

WEATHER
CHAPTER 6:
Weather Graphs

What month is the hottest where you live? The rainiest? How does where you live compare to other cities around the world?

Graphs can be used to show lots of data in a way that makes it easy to answer questions like the ones above.

By the end of this chapter, you should know how to read pictographs and bar graphs that display weather data, and even make some graphs of your own.

―――――――――――――――――――――――――
 = Comic
―――――――――――――――――――――――――

CHAPTER 6:
WEATHER GRAPHS

We can collect weather data to look for patterns, but patterns are often difficult to see with just words and numbers.

Graphs can make it easier to see patterns in data, and can help us explain these patterns to others.

Where have you seen graphs used to show data?

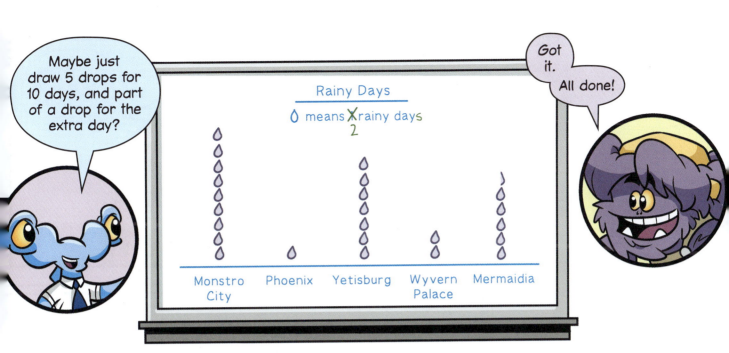

111

PICTOGRAPHS

A **pictograph** is a way of showing data with pictures or symbols. Each picture stands for some number of things. For example, in the pictograph below, each sun stands for 20 sunny days (days when the sky is clear or mostly clear all day).

The graph shows about how many sunny days there are in six U.S. cities.

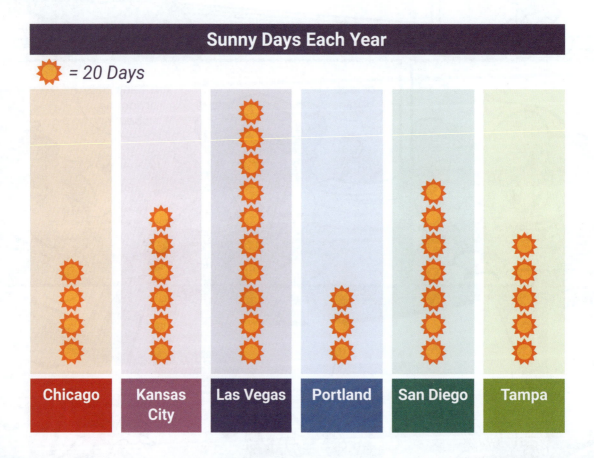

1   Which city above has the fewest sunny days each year? Which has the most?

2   About how many more sunny days are there in Las Vegas each year than in San Diego?

Use the pictograph on the previous page to answer the questions below.

3 Mark all of the statements below that are true in a normal year.

○   **Chicago has more sunny days than Portland.**

○   **San Diego has about 20 more sunny days than Kansas City.**

○   **Las Vegas has about twice as many sunny days as Tampa.**

4 Sometimes, pictographs use partial symbols like ☀. How many sunny days would a ☀ stand for in the pictograph on page 114?

Sunny Days

5 People in San Diego get about as many sunny days each year as which two cities combined?

                       +

6 If a ☀ stands for 20 sunny days, it's hard to show exactly 84 sunny days on a graph. Below is the actual data used to make the pictograph on page 114. How many sunny days are **missing** from the pictograph for each city?

Sunny Days Each Year

City	Chicago	Kansas City	Las Vegas	Portland	San Diego	Tampa
Sunny Days	84	120	209	68	146	101
Missing Days						

BAD PICTOGRAPHS ⚠️

Some of the pictographs on these pages are not as clear as they could be. Think about ways you could make each graph better as you answer the questions below.

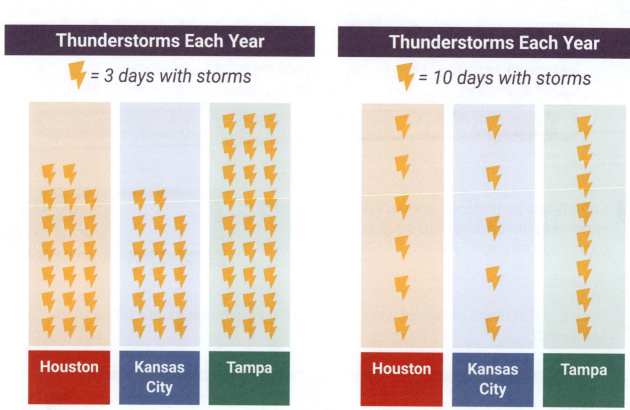

7 What would you change about the pictograph on the left to make it easier to tell how many days have thunderstorms in each city?

_____

_____

8 What would you change about the pictograph on the right to make it easier to tell how many more thunderstorms there are in Tampa than in Houston each year?

_____

_____

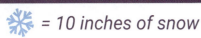
Snowfall Each Year
❄ = 10 inches of snow

| Boston | Chicago | New York |

Snowfall Each Year
❄ = 10 inches of snow

| Boston | Chicago | New York |

9 Which of the three cities receives the most snow? Which graph makes it easier to tell? Explain.

_____

_____

10 What should be true of the pictures used in a pictograph where each picture stands for the same amount? Mark all that apply.

⚪ All of the pictures should be the same size.

⚪ All of the pictures should be spaced the same way.

⚪ Every stack of pictures should be the same height.

11 Of the four pictographs on pages 116 and 117, which do you think does the best job of showing information? Explain.

_____

_____

BAR GRAPHS

A **bar graph** looks a lot like a pictograph. Instead of pictures, a bar graph uses bars of different lengths to stand for amounts. The taller the bar, the greater the amount. The left side of a bar graph usually shows the amounts that the different bar heights stand for.

Below, we show normal daily high temperatures in Seattle and Chicago using both a table and a bar graph.

Normal Daily High Temperatures by Month (°F)												
	Jan	Feb	Mar	Apr	May	Jun	Jul	Aug	Sep	Oct	Nov	Dec
Seattle	47°	49°	52°	57°	63°	66°	72°	73°	67°	59°	51°	46°
Chicago	32°	36°	45°	56°	66°	77°	82°	81°	74°	62°	50°	37°

DISCUSSION:

12 What patterns do you notice in the data above? Is it easier to see these patterns by looking at the table or the bar graph?

Use the table and bar graph on the previous page to answer the questions below about the normal daily high temperatures in Seattle and Chicago.

13 What is the normal daily high temperature in Seattle for each month listed below?

March   June   September   December

14 Based on the normal daily high temperatures, which city has the hottest month? Which city has the coldest month?

Hottest   Coldest

15 Based on the normal daily high temperatures, what is the hottest month for each city?

Seattle   Chicago

16 Which city has bigger temperature changes during the year? Explain.

_____

_____

17 Which of the questions above was easier to answer by looking at the table? Which was easier to answer with the graph? When did you use both?

MISLEADING BAR GRAPHS ⚠️

Graphs are great at showing patterns and relationships, but they can sometimes be **misleading**. A misleading graph suggests something that is not true.

PRACTICE:

Both graphs below show the same data for days when it's mostly cloudy and mostly sunny in Denver. Use the graphs to answer the questions that follow.

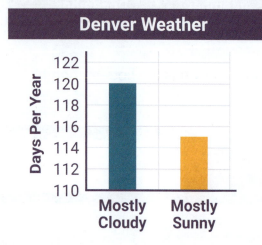

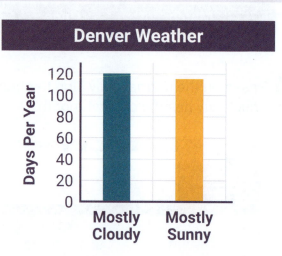

18 According to both graphs above, about how many mostly cloudy days and mostly sunny days are there in Denver every year?

Mostly Cloudy

Mostly Sunny

19 Mark the true statement below.

⭕ **Denver has about the same number of cloudy and sunny days each year.**

⭕ **Denver has about twice as many cloudy days as sunny days each year.**

DISCUSSION:

20 Describe how one of the graphs above might be confusing.

Answer the questions about the bar graphs below.

21 Circle the graph on the right that suggests that it rains about twice as much in Dallas as it does in San Jose. Then, explain whether you think the graph you circled is misleading or not.

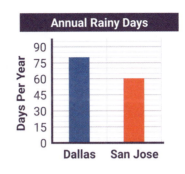

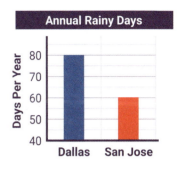

_____

_____

22 Circle the graph on the right that does a better job of showing the changes in Miami's normal high temperatures through the year. Explain why.

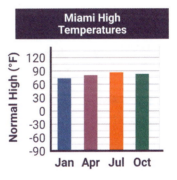

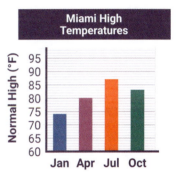

_____

_____

23 The graph on the right compares the snowfall at two major ski resorts. About how many inches of snow fall at each resort? How is this graph misleading?

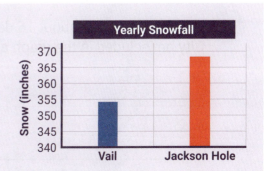

MAKE A BAR GRAPH

Redraw a misleading bar graph.

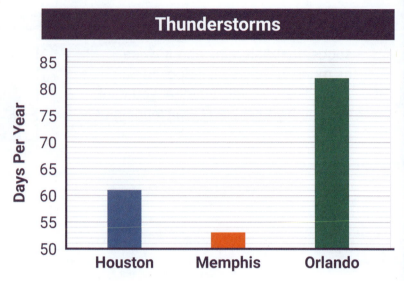

24 About how many days does it storm each year in the cities below?

Houston ☐   Memphis ☐   Orlando ☐

25 Your friend sees the graph above and says it almost never storms in Memphis. Are they right? Explain.

_____

_____

26 New York City has about 15 days with thunderstorms every year. Why can't New York City be included in the graph above?

_____

_____

Draw a new graph that compares the number of days when there are thunderstorms in Houston, Memphis, and Orlando.

27 Fill in a bar on the graph below for each city to show about how many days have thunderstorms each year.

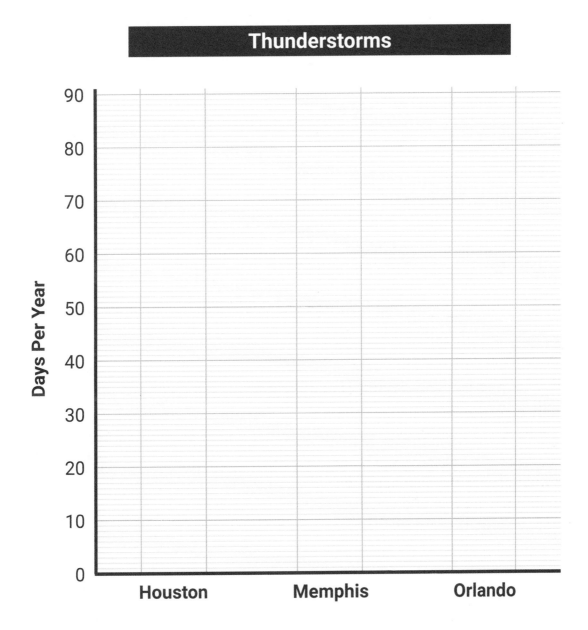

Thunderstorms

Days Per Year: 90, 80, 70, 60, 50, 40, 30, 20, 10, 0

Houston   Memphis   Orlando

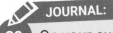

JOURNAL:

28 On your own graph paper, make a bar graph for the numbers of rainy days or yearly rainfall amounts for any two of the cities on pages 96-97, plus your own city.

WEATHER
CHAPTER 7:
Day Length

Why does it get dark early in some months, but stay light late in others?

Are there places where it can stay light or dark for days at a time, or even months?

By the end of this chapter, you should know what causes the changes in day length through the year and understand some of the effects of these changes.

=Lab   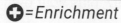=Enrichment

CHAPTER 7:
DAY LENGTH

Have you ever noticed that the sunrise and sunset times change from month to month?

Graphs can help us see how these changes in day length take place in different parts of the world.

The famous clock tower at the Palace of Westminster in London is nicknamed Big Ben.

The two photos of Big Ben were taken after sunset, but before the sky is completely dark.

Look at the times in the photos. In what part of the year do you think each picture was taken?

SEASONS

Seasons are parts of the year that have different weather patterns. In many parts of the world, each year is divided into four seasons: spring, summer, fall, and winter.

One of the most important changes from season to season is the length of each day. Unless you live very close to the equator, the amount of time when the Sun is up changes from month to month.

Below we graph the day length in New York City in the middle of each month.

New York City
USA
North America
South America

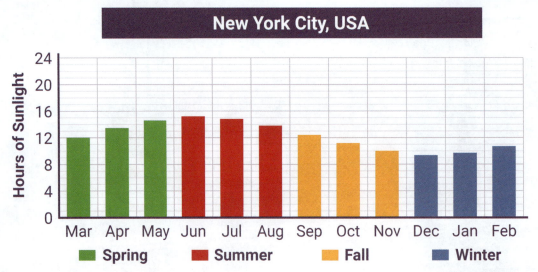

New York City, USA

DISCUSSION:

1 What patterns do you see in the day lengths above?

When do the seasons start and end?

In the graph above, each season lasts for three months. This is how many meteorologists (people who study the weather) define the seasons.

Astronomers (people who study objects in space) define seasons differently. Summer begins on the longest day of the year and winter begins on the shortest day of the year. Spring and fall begin when days are about 12 hours long.

Answer the questions below using the graph on the previous page.

2 What month has the longest days in New York City? _____

3 What month has the shortest days in New York City? _____

4 During which two months are the days closest to 12 hours long in New York City?

_____   _____

DISCUSSION:

5 Look at the sunlight graph for Santiago, Chile, below. What is different between the seasons in Santiago and New York City? Are your answers to the questions above the same for Santiago as they are for New York City?

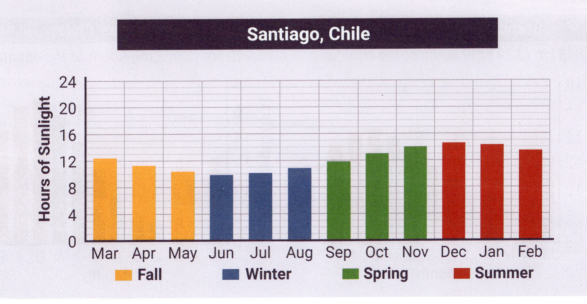

AROUND THE GLOBE

On the previous page we saw that in months when New York City days are longest, Santiago is experiencing its shortest days.

New York City is north of the equator, while Santiago is south of the equator.

Look for patterns in the graphs below to learn more about how location on the globe affects day length around the world.

Montreal, Canada
5,064 km (3,147 mi) north of the equator

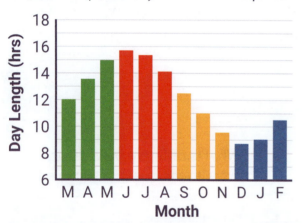

Havana, Cuba
2,572 km (1,598 mi) north of the equator

Buenos Aires, Argentina
3,848 km (2,391 mi) south of the equator

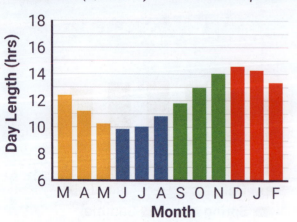

Punta Arenas, Chile
5,909 km (3,672 mi) south of the equator

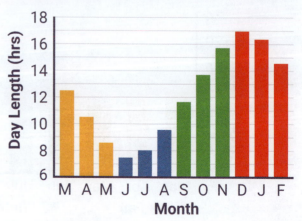

Answer the questions below about the graphs on the previous page.

6 Which two cities have their longest days in December? _____

7 Which city **never** has a day longer than 14 hours? _____

8 List the cities in order from the city with the longest summer days at the top to the city with the shortest summer days at the bottom.

_____

_____

Remember, summer months are different north and south of the equator.

_____

_____

9 Which city has the **largest** difference between its longest days and its shortest days? _____

10 Which city has the **smallest** difference between its longest days and its shortest days? _____

JOURNAL:

11 Look at the distance from each city to the equator.

Describe the relationship you see in the graphs between distance from the equator and the differences in day length through the year.

EARTH AND THE SUN

To understand what causes the changing day lengths, we need to understand how Earth moves around the Sun.

Every day, Earth spins once around its axis—an imaginary line that connects the North Pole and the South Pole through the middle of the planet.

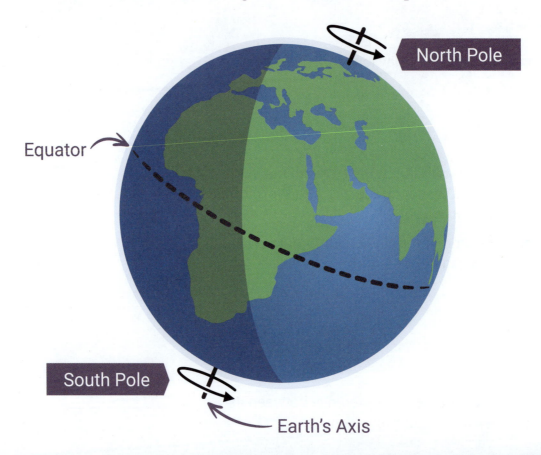

Equator

North Pole

South Pole

Earth's Axis

Important! How big and far apart are Earth and the Sun?

None of the diagrams in this chapter are drawn to scale. In other words, things are not the right size compared to each other. If Earth were the size it is in the drawing above, the Sun would be about 10 meters tall and 1 kilometer away!

To put Earth and the Sun together on a page in this book, Earth has to be too small to see!

Earth
(Way too small to see. It would fit on the tip of a single hair.)

LAPS AROUND THE SUN

Every year, Earth makes one big lap around the Sun.
Earth's axis is tilted as shown below. So, different parts of Earth face the Sun at different times of the year. In June, the North Pole is tilted toward the Sun. In December, the North Pole is tilted away from the Sun.

The tilt causes changes in day lengths through the year as Earth makes its lap around the Sun.

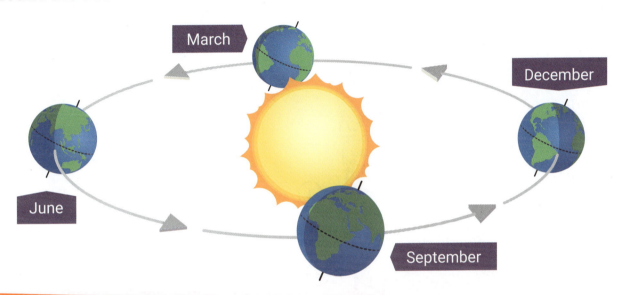

March

December

June

September

DOES THE DISTANCE FROM EARTH TO THE SUN CHANGE?

The distance from Earth to the Sun does change a little during the year, but this doesn't cause the seasons. Earth is closest to the Sun in early January and farthest in early July.

When you look at the diagrams in this chapter, it's important to remember that Earth is much smaller than the Sun, and very far away. The Sun is more than 100 Earths wide, and you could fit more than 100 Suns between Earth and the Sun.

The Sun

MAKE A GLOBE

How does Earth's tilt change the way sunlight hits our planet?
Make a model of Earth and pretend you're the Sun to find out.

(If you already have a globe, you can use it instead.)

MATERIALS

You will need:

- One orange
  (or a similar round fruit or
  foam craft ball)
- Permanent marker
- Sharp pencil
- Rubber band
- Headlamp (optional)

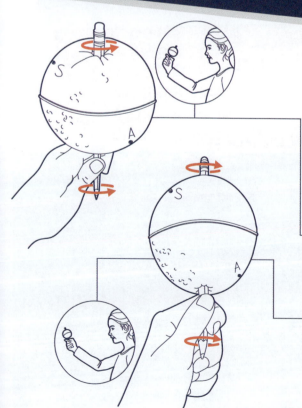

1. Your orange will be your "globe". Poke a pencil through it to make its axis and stretch a rubber band around it as the equator.

2. Add dots north and south of the equator and label them "Seville" and "Auckland" or just "S" and "A" as shown below. They should be opposite each other on your orange.

Pretend your head is the Sun!

3. If you have a headlamp, put it on and turn the lights off to make this demonstration more dramatic! Hold the orange out in front of you and watch how what you see changes as you tilt and turn your little model of our planet.

4. Tilt the orange toward you as shown. This is what Earth looks like from the Sun in **June**.

   Slowly spin the orange.

5. Tilt the orange away from you as shown. This is what Earth looks like from the Sun in **December**.

   Slowly spin the orange.

Answer the questions below about your orange "globe" model of Earth.

12 In June, do you see more of the top half or the bottom half of your orange planet? _____

13 In June, as you spin your orange one time around (a full turn is one whole day on your little orange planet), which dot do you see for more of each day: Seville or Auckland? _____

14 In June, are there parts of the orange that you never see as you turn it? If so, what parts?

_____

_____

15 How would your answers to the questions above change for December instead of June?

Where does that hole go?

Auckland and Seville are on opposite sides of the planet. If you could dig a hole straight through the center of Earth from Auckland, New Zealand, you would end up near Seville, Spain!

Places on opposite sides of the globe are called **antipodes**. Where would you end up if you dug straight down from where you live? Search online for "antipode" and a nearby city to find out. Most people would end up in an ocean. Earth is covered by more water than land, after all.

DAY LENGTH

In June, the North Pole is tilted toward the Sun. Think about how this affects day lengths for places north and south of the equator as Earth spins.

16 This is the path of New York City as Earth turns in June. Does New York City spend more time in the light or the dark in June?

June Sunlight

17 This is the path of a research station on the coast of Antarctica as Earth turns in June. What do you think days are like at this research station in June?

In December, the North Pole is tilted away from the Sun. Think about how this affects day lengths for places north and south of the equator as Earth spins.

18 This is the path of New York City as Earth turns in December. Does New York City spend more time in the light or the dark in December?

_____

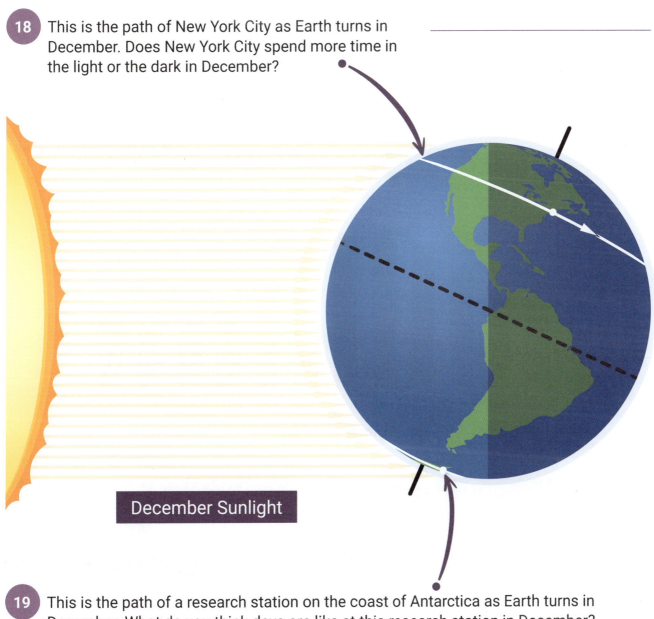

December Sunlight

19 This is the path of a research station on the coast of Antarctica as Earth turns in December. What do you think days are like at this research station in December?

_____

_____

SEASONAL TEMPERATURES

In places that have four seasons, temperature is the most obvious seasonal weather change. Temperatures change throughout the year.

Below, we graph the normal daily high temperatures in Rome for each month.

What patterns do you notice?

Rome, Italy

Asia

Africa

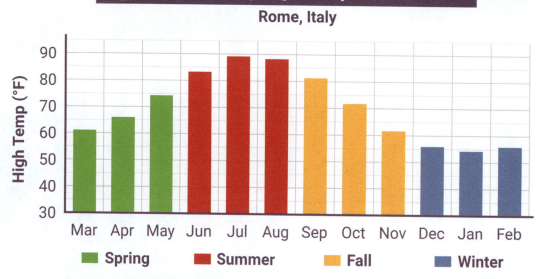

Normal Daily High Temperatures
Rome, Italy

High Temp (°F)

- ■ Spring
- ■ Summer
- ■ Fall
- ■ Winter

Why aren't the longest days the hottest?

North of the equator, the longest day of the year is usually June 21st. You might think that the longest day of the year would be the hottest. But in cities like Rome, Atlanta, and Tokyo, the temperatures are hottest in July and August. Why?

It's because things take a while to heat up, especially water. Earth's surface is like a giant pot of water that takes a long time to heat up (and cool down).

Answer the questions below using the graph on the previous page.

20 What season has the highest temperatures in Rome? _____

21 What season has the lowest temperatures in Rome? _____

22 During which season do temperatures increase every month in Rome? _____

23 Look at the temperature graph for Sydney below. What is different between the seasons in Sydney and Rome? Are your answers to the three questions above the same for Sydney as they are for Rome?

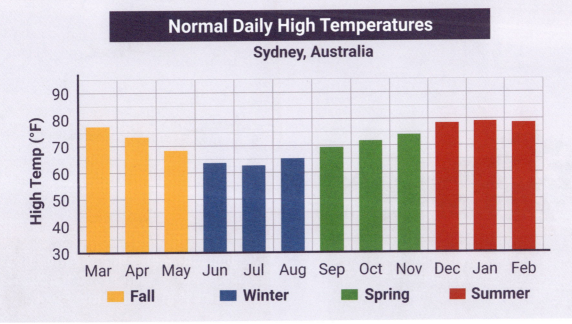

Normal Daily High Temperatures
Sydney, Australia

THE POLES

The North Pole and South Pole have cold summers and extremely cold winters.

People don't live year-round at the poles. But, in parts of Alaska, Scandinavia, Canada, and Russia, people live in places where the Sun stays up for months in summer and doesn't rise for months in winter.

These areas are called the **polar zones**.

The Polar Zone around the North Pole is called the **Arctic**.

Ilulissat, Greenland
About 5,000 people live in Greenland's third-largest city.

The Polar Zone around the South Pole is called the **Antarctic**.

Amundsen-Scott Station, Antarctica
This research station in Antarctica is located at the South Pole!

Use the sunlight graph below to help you answer the questions that follow.

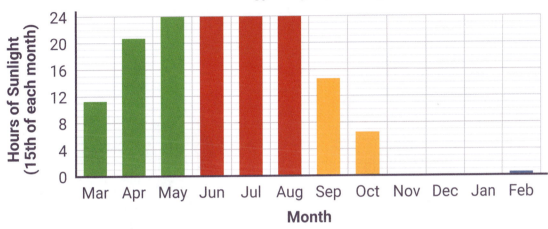

24 There are four months in Longyearbyen when the Sun stays up all day on the 15th. List them. (You can just write the first three letters of each month as in the graph.)

_____   _____

_____   _____

25 After months without sunlight, during what month does the Sun finally rise in Longyearbyen? _____

26 Longyearben is near the North Pole. What months do you think have sun all day near the South Pole?

_____

_____

THE TROPICS

The area near the equator is called the **tropical zone** or just **the tropics**.

The tropical zone includes areas with lots of different weather patterns: rainy jungles, dry deserts, warm beaches, and even snow-capped mountains. What these areas all have in common is that the day lengths don't change much from month to month.

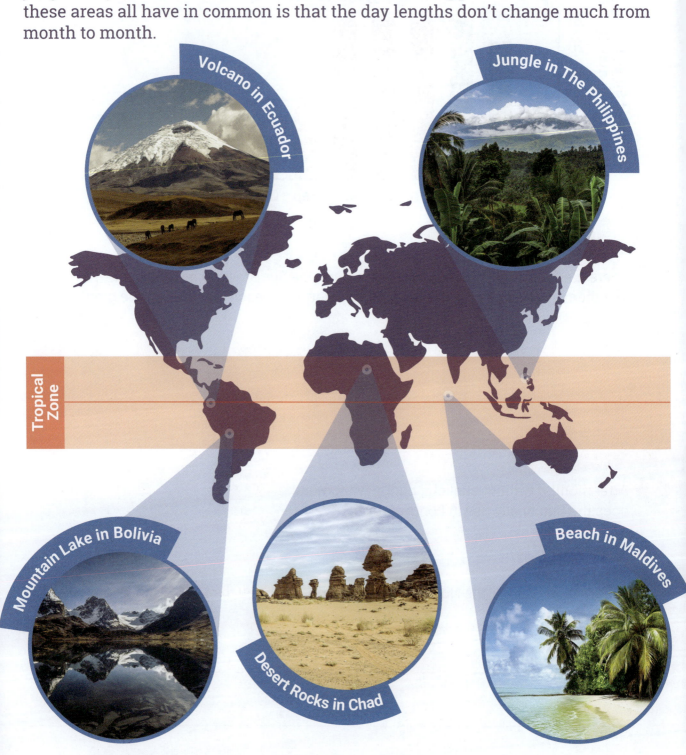

Volcano in Ecuador

Jungle in The Philippines

Tropical Zone

Mountain Lake in Bolivia

Desert Rocks in Chad

Beach in Maldives

DISCUSSION:

27 Singapore is very close to the equator. Based on the graph below, do you think people in Singapore experience four seasons of spring, summer, fall, and winter the way many people do in other parts of the world?

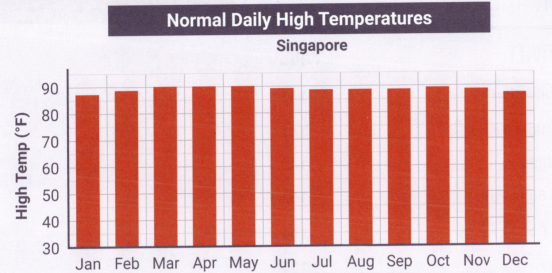

28 Which of these could describe a place in the tropics? (Check all that apply.)

○ A mountain town where you may need a sweater all year.

○ A desert village where it doesn't rain for months at a time.

○ A city where the Sun doesn't set for months.

○ An island where temperatures are almost always hot.

29 Describe one thing that you think would be better about living in the tropics than near the poles, and one thing that you think would be better about living near the poles than in the tropics.

_____

_____

WEATHER
CHAPTER 8:
Sun Angle

How does the Sun move across the sky in different parts of the world? Where and when does it get highest in the sky? Where and when does it stay low near the horizon? And why is your shadow longer or shorter at different times of day?

By the end of this chapter, you should understand how the angle of the sunlight that reaches Earth's surface affects different parts of the globe.

🧪 = Lab   ➕ = Enrichment

CHAPTER 8: SUN ANGLE

Day length is just part of the reason for seasonal temperature changes around the world.

If day length was all that mattered, the poles would be very hot during months when the Sun shines 24 hours a day!

Sun angle makes a big difference in how much heat and light reaches Earth's surface.

Sunbeams appear to meet above the clouds for the same reason that train tracks appear to come together in the distance. Things that are far away look smaller and closer together than when they are close. But, like the tracks of a train, these sunbeams are all parallel—they are all going the same direction.

On the right, we can see what the shadows of clouds and the sunbeams between them look like from high above. This photo was taken by an astronaut aboard the International Space Station.

SPOTLIGHT DEMO

The angle of the Sun—how high the Sun is above the horizon—changes during the day and during each season.

When does the Sun provide the most heat and light to Earth's surface?

MATERIALS

You will need:

- Flashlight
- Dark room
- White sheet of paper
- Scissors

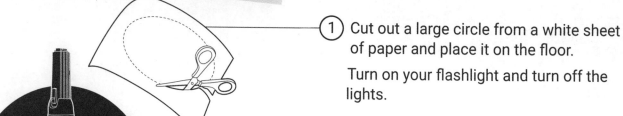

(1) Cut out a large circle from a white sheet of paper and place it on the floor.

Turn on your flashlight and turn off the lights.

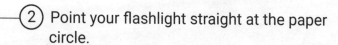

(2) Point your flashlight straight at the paper circle.

Move it close enough so that light fills the circle.

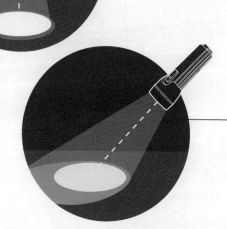

(3) Next, tilt your flashlight. Keep it aimed at the center of the circle and about the same distance away as it was before.

Answer the questions below about the sun-angle demo.

1 Does the paper circle look brighter when the flashlight is shining from straight above (as in step 2) or when the light is shining at a lower angle (as in step 3?)

_____

2 Does the light spread over more area when the flashlight is straight above or when the light is at an angle?

_____

3 About what time of day is the Sun highest in the sky where you live? Morning? Lunchtime? Evening? When do you think it provides the most heat and light?

_____

_____

DISCUSSION:

4 Earth is round. So, the angle that the Sun hits Earth's surface changes depending on where you are. Where do you think the Sun feels brightest and hottest on the surface of the globe below?

SUNLIGHT

In June, the North Pole is tilted toward the Sun. Think about how this affects sunlight on different parts of the globe as Earth spins.

5 The colored regions A, B, C, and D are the same size, but get different amounts of sun in June. List the regions in order from the region that gets the least sunlight to the region that gets the most.

1 _____ least   2 _____

3 _____   4 _____ most

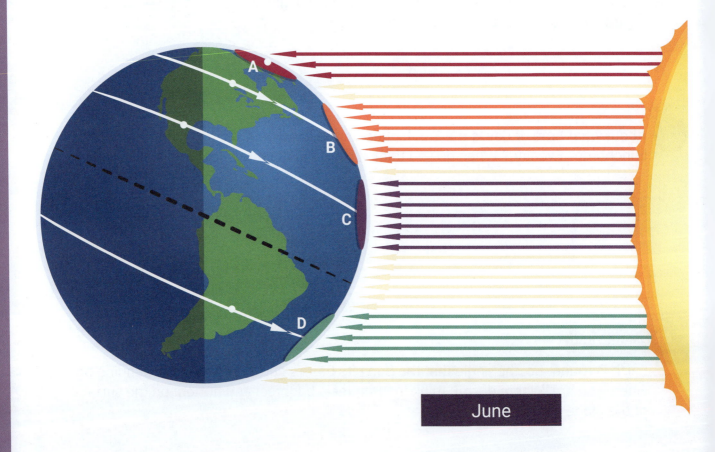

June

DISCUSSION:

6 The white dots mark locations that pass through each region: The North Pole is in region A. Winnipeg, Canada passes through B. Monterrey, Mexico passes through C. Asuncion, Paraguay passes through D.

The North Pole has 24-hour sunlight in June, but it is the coldest of these four locations. Why do you think the North Pole is coldest?

In December, the North Pole is tilted away from the Sun. Think about how this affects sunlight for places north and south of the equator as Earth spins.

7 Which region below gets the same amount of sun in December as region C gets in June? _____

8 Which region has the greatest difference in sun between June and December? (Count the "rays" that strike each region and subtract.) _____

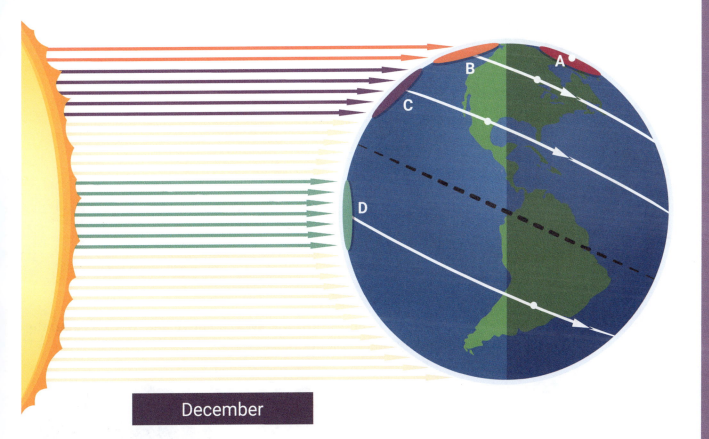

December

9 Which region above gets as much sun in December as all three of the other regions combined? _____

💬 DISCUSSION:

10 Imagine a region E that is the same size as regions A, B, C, and D, but centered on the equator between regions C and D. How would the amount of sun it gets compare to regions C and D in June? In December?

SUN ANGLE

Earth's tilt changes the way the Sun shines on different parts of the planet as it moves around the Sun.

When the Sun is directly above you, it delivers more heat and light than when it is closer to the horizon.

Imagine what the Sun looks like to a person standing on Earth near the poles and to a person near the equator. (Never look directly at the Sun.)

11 Label the images below A, B, and C to show how the people above see (or don't see) the Sun from where they are standing on the planet.

Answer the questions below about how the Sun looks in different parts of the year.

12 Shade the section of the globe where the Sun will get highest in the sky as Earth spins in March. Think about how a person standing on Earth sees the Sun.

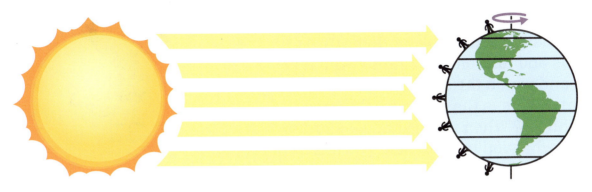

13 Shade the section of the globe where the Sun will get highest in the sky as Earth spins in June.

14 Shade the section of the globe that will not get any sun as Earth spins in December.

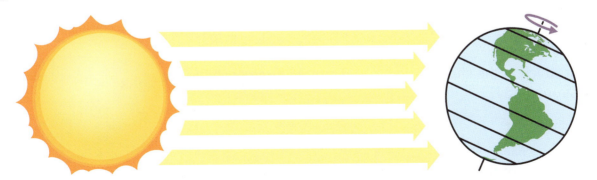

SHADOWS

You shouldn't look at the Sun, but there's an easy way to tell where it is. You can look at your shadow! If you could draw an arrow from your shadow's head through your head, it would point right at the Sun!

Place a small toy on a table or floor. Turn off the lights and turn on a flashlight. Aim it at the toy and watch the shadow as you move the light. Where is the shadow compared to the flashlight? When is the shadow short or tall? Use what you learn to help you answer the questions below.

PRACTICE:

Answer the questions about the shadows of the astronaut toy below.

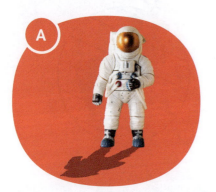
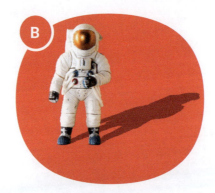
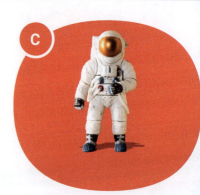

15 For which of the shadows is the Sun **behind** our astronaut toy? _____

16 For which of the shadows is the Sun **almost directly above** our astronaut toy? _____

17 For which of the shadows is the Sun **lowest** in the sky? _____

Answer the questions below about shadows.

18 This photo shows an overhead view of players on a field. Draw an arrow from the dot in the middle of the field that points toward the Sun.

19 About what time of day was the photo on the right taken?

- ⬤ **Early Morning**
- ⬤ **Sunset**
- ⬤ **Noon**
- ⬤ **Late Afternoon**

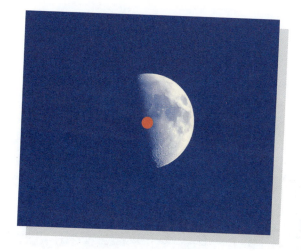

20 Like Earth, the Moon is a big sphere that is lit on one side by the Sun. The other side is in shadow. So, the light side of the moon lets us know where the Sun is (even at night).

Draw an arrow from the dot on the Moon that points toward the Sun.

THE SUN'S PATH

From Earth's surface, the Sun seems to move across the sky from east to west. This is because Earth is spinning, not because the Sun is moving around our planet. The Sun's path changes during the year. Here, we see how the Sun's path looks at different times of year in Vancouver, Canada.

When the pole you are closest to is tilted toward the Sun, the Sun gets higher and stays in the sky longer than when the pole is tilted away from the Sun.

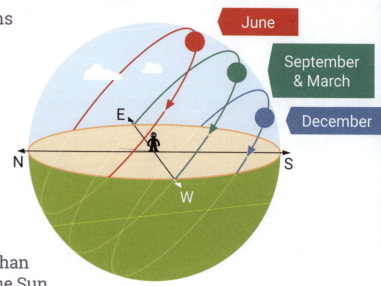

June

September & March

December

21 Write the months June, September, and December in the blanks to show what the path of the Sun looks like to a person north of the equator looking south.

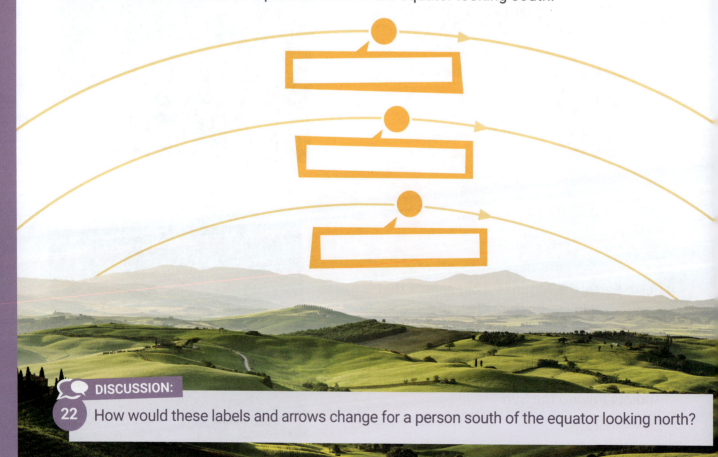

DISCUSSION:

22 How would these labels and arrows change for a person south of the equator looking north?

ACROSS THE SKY

The Sun isn't the only thing that seems to move as Earth turns.

For thousands of years, people have watched the stars and tracked their movements. A great way to see their paths is with a star trail image of the night sky.

Imagine taking a photo every 20 seconds for an hour or more, then stacking them all into one image. That's the idea behind these star trail images that show the paths of stars across the night sky.

DISCUSSION:

23 What do you think made the bright streaks in the images on the right and below?

SUN BOX

Build a box that lets you record the path of the Sun in the sky where you live.

MATERIALS

You will need:

- Shoe box
- Dark construction paper
- Scissors and tape
- Adult help
  (with a knife or box cutter)

MAKE A SUN 'CAMERA'

Long before there were digital cameras, special paper that is sensitive to light was used to make photos. Even normal construction paper has dyes that fade in sunlight.

In this project, we'll build a box that works a lot like a camera. We'll use dark construction paper as our 'photo' paper to capture an image of the Sun's path.

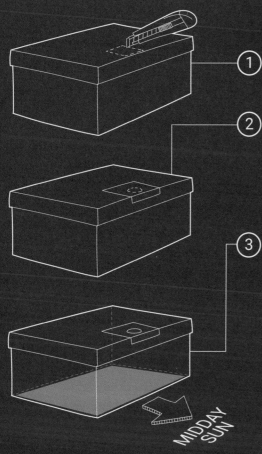

1. Ask an adult to cut a 3-4 cm square hole in the middle of the top-front edge of a shoe box as shown.

2. Cut a coin-sized (2 cm) round hole in a piece of construction paper and tape it over the hole in the box. Cover any other holes in the box with construction paper so that light only enters the box through the round hole.

3. Cover the bottom of the inside of the shoe box with a dark piece of construction paper.

In the morning on a sunny day, place the box where it will get sun all day. The hole in the box should be pointed toward the midday sun (south if you live north of the equator, north if you live south of the equator).

At the end of the day, open the box and look at the construction paper inside.

Depending on where you live and the time of year, the sun may not be strong enough to fade the paper in one day. It may help to leave the box out for several days, but be careful not to let rain ruin your sun box.

 JOURNAL:

24. Tape the faded construction paper to a page in your science journal. Describe what you notice about how the paper faded. Is the paper faded in some places more than others?

 TIP:

Put something on your box that will keep it from moving or blowing away, but be careful not to block the hole from letting the sunlight in.

(EXTENSION) Waterproof your box and leave your box in the sun for a month or more. You can use a clear plastic bag to waterproof your sun box, but make sure sun still shines through the hole. What does your construction paper look like after a month?

25 What are the *two* main reasons summer is hotter than winter?

- ○ The Sun gets higher in the sky in summer than in winter.
- ○ Earth is closer to the Sun in summer than in winter.
- ○ Summer days are longer than winter days.
- ○ The Sun is larger in summer than in winter.

26 What is the main explanation for the two reasons you chose above?

- ○ Earth's path is not a circle.
- ○ Earth is tilted in relation to the Sun.
- ○ The Sun shrinks and grows.
- ○ Earth wobbles as it spins.

27 South of the equator, what month has the longest days and the highest Sun angle?

- ○ March
- ○ June
- ○ September
- ○ December

28 If the Sun had eyes and could look all the way to Earth, could it see the shadows of clouds and skyscrapers? Explain.

_____

_____

_____

29 During what season and time of day will your shadow be shortest outdoors? Explain.

_____

_____

30 Solar panels collect energy from the Sun. Why are most solar panels tilted instead of pointed straight up?

_____

_____

_____

31 There is a place on Earth where the Sun doesn't always rise in the east and set in the west.

Instead, the Sun rises in March and traces a circle around the horizon. It slowly spirals upwards until June.

Then, the Sun circles back down until September when it drops below the horizon again.

Where is this place?

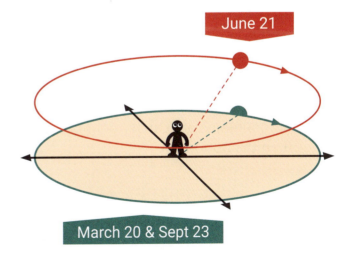

June 21

March 20 & Sept 23

_____

WEATHER
CHAPTER 9:
The Atmosphere

Are there cold places near the equator?

How can two places that are only a few miles apart have very different temperatures?

By the end of this chapter, you should understand more about how the air all around us gets heated, and why the tops of mountains are usually so much colder than the bottoms.

🧪=Lab   AHA!=Comic   ➕=Enrichment

CHAPTER 9:
THE ATMOSPHERE

Weather describes the conditions in our **atmosphere**—the air all around us. Distance from the equator is not the only thing that affects weather patterns. How does the atmosphere change as you get farther from Earth's surface?

Mount Kilimanjaro is the highest mountain in Africa, rising 5,895 meters (19,341 ft) above sea level in Tanzania, near the Kenyan border.

It is practically on the equator, so temperatures on the mountain don't change much from month to month. But, temperatures change a lot between the base of the mountain and its peak.

Its peak stays below freezing and has ice and snow year-round. At the base of the mountain, temperatures are usually in the 70's and 80's. Why?

What keeps lots of things from getting crushed by air... is *air!*
Huh?
I'll show you.
Try to crush this bottle with your hands.
No problem, Dr. Hau.
I can't do it.
It keeps springing back.
Why?
It's full of air.
I could crush it if I take the cap off.
You don't even need your hands to crush it with the cap off.

163

Maybe it was the invisible superhero?

Nope.

Was it... ...the atmosphere?

It was! Just a little air doesn't weigh much, but there is a *lot* of air above us.

The atmosphere is deeper than the tallest mountains.

All that air weighs a lot!

The weight of the atmosphere is always pushing on the surfaces all around us.

How hard the air pushes is called *air pressure*.

Air doesn't *feel* heavy.

Why can't I feel the weight of all the air above me?

AN OCEAN OF AIR

Air doesn't weigh much, but it probably weighs more than you think. A typical bedroom holds about 40 kilograms (around 90 pounds) of air!

Earth's atmosphere is like a really deep ocean of air that covers Earth's surface. The weight of all that air is constantly pressing on everything, just like water does in a pool. If you've ever gone deep under water, you've probably felt the pressure in your ears. The deeper you go, the more pressure you feel.

The atmosphere works in a similar way, pressing on us from all sides. The amount that air presses against things is called **air pressure**.

PRACTICE:

Understanding air pressure is difficult because we can't see air.
Think about how things you can see work when they are piled on top of each other.

1 When building a human tower, does everyone in the tower carry the same amount of weight? If not, which people have the most pressure on their shoulders?

_____

_____

2 Each of these boxes is completely filled with marbles. If you could stack more boxes on top without tipping them over, which box do you think would break first? Why?

_____

_____

_____

AIR PRESSURE IN THE ATMOSPHERE

Air is invisible, so it's hard to imagine how air is different in different parts of the atmosphere. So, we'll imagine the next best thing—marshmallows!

These tall tubes are filled with colorful candies. Imagine a strong clear tube 30 feet high (about 10 meters) filled with marshmallows instead.

Marshmallows are light because they are mostly air, but each marshmallow at the bottom of the tube would have over 1 kilogram (more than 2 pounds) of marshmallows directly above it. That's about the weight of a baseball bat.

3 How do you think the marshmallows near the bottom of the tube would look different from the marshmallows near the top?

_____

_____

_____

 DISCUSSION:

4 Imagine you are a ping pong ball near the bottom of the tube, surrounded by marshmallows. Do you think it would feel like you were being squeezed by the marshmallows from above, from below, from the sides, or from everywhere all at once? Explain.

MARSHMALLOW BOTTLE

Is air really squeezing everything all around us?

How can we see the effects of air pressure if air is invisible?

MATERIALS

You will need:

- Air pump with an inflating needle
- Clear plastic bottle
- A hammer and a nail that is the same size or slightly smaller than the inflating needle
- Marshmallows (The more, the better.)

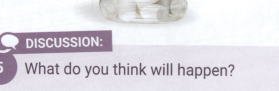

With help from an adult:

1. Remove the cap from the bottle and add marshmallows. The more, the better. If your bottle has a small opening, you may need to use mini marshmallows.

2. Hammer the nail through the center of the cap to create a hole that is just barely big enough to push the inflating needle through. Screw the cap tightly back on the bottle.

3. Push the inflating needle through the hole in the cap and start pumping.

4. When it gets difficult to pump, remove the needle from the cap to let the air escape from the bottle.

TIP: *Record a video of the bottle so you can watch the changes sped up.*

DISCUSSION:

5. What do you think will happen?

Answer the questions below about what happened to the marshmallows in the bottle.

6 What happens to the size of the marshmallows as air is pumped into the bottle?

_____

_____

7 What happens to the size of the marshmallows when air is released from the bottle?

_____

_____

8 What happens to the shape of the marshmallows as air is pumped in and released? (Are they flattened like pancakes? What shape are they when the pressure is greatest?)

_____

_____

DISCUSSION:

9 What do you think would happen to the marshmallows if you could pump air **out** of the bottle? (You can try sucking air out through the tiny hole. It's not easy to suck enough air out to see significant changes, but if you film the marshmallows, you may be able to see them change.)

ALTITUDE

Standing at the base of a mountain, you are near the bottom of the deep ocean of air we call the atmosphere. As you gain **altitude** (height above Earth's surface), you get closer to the top of the atmosphere. When there is less air above you, the air pressure decreases.

AIR

The higher you go, the lower the air pressure gets.

You may have driven up a mountain and felt your ears 'pop'. This happens when the pressure of the air behind your eardrum is higher than the pressure of the air around you.

Pressure squeezes air so that it takes up less space, just like the marshmallows in the demo.

10 An altimeter is an instrument in a plane that displays altitude by measuring air pressure. How do you think the altitude shown on an altimeter changes as the pressure changes?

- ○ **As air pressure drops, altitude increases.**

- ○ **As air pressure drops, altitude decreases.**

- ○ **As air pressure drops, altitude does not change.**

DISCUSSION:

11 What do you think would happen to the reading on an altimeter if you placed it inside the marshmallow bottle on page 168 and pumped it up?

DISCUSSION:

12 This bag of chips looked normal when it was bought at the bottom of a mountain, but on the drive up the mountain, it puffed up until it almost popped. If no air could get in or out of the bag, why do you think this happened?

AFTER

BEFORE

ALTITUDE AND TEMPERATURE

If you have ever been high up on a mountain, you probably noticed that the top is colder than the bottom.

It's not just the mountain that's colder up high, it's the whole atmosphere!

As you move up through the atmosphere, the air temperature usually drops about 1°C for every 100 meters you climb (about 1°F for every 180 feet). To understand why, we need to understand how pressure affects temperature.

In winter, Los Angeles usually has high temperatures around 68°F (20°C). Temperatures there almost never get below freezing.

But, the nearby mountain peaks are often covered in snow. This is because the air up high is almost always colder than the air down low.

UNDER PRESSURE

How are air pressure and temperature related?
Does air cool down under pressure? Or, does air heat up when it's squeezed?

Pump up a bicycle tire to find out.

MATERIALS

You will need:

- Bicycle tire
  (any tire you can inflate
  with a bicycle pump)

- Bicycle pump

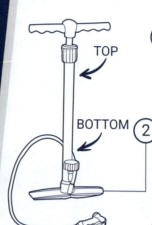

TOP

BOTTOM

With help from an adult:

1. Let the air out of an inflated bicycle tire. Pay attention to the temperature of the air that comes out of the tire.

2. Now, pump the tire back up. As one person uses the bicycle pump to inflate the tire, the other should feel the temperature at the top and the bottom of the bicycle pump's tube.

PRACTICE:

Answer the questions below about the temperatures you felt.

13. Does the air that comes out of the bicycle tire feel colder, warmer, or about the same temperature as the surrounding air? (Make sure the tire has not been stored somewhere cold or hot.)

_____

14. Does the temperature of the bottom of the bicycle pump tube (where pressure gets highest) feel colder, warmer, or about the same temperature as at the top of the tube (where there is less pressure)?

_____

15 It's time to bring back your marshmallow bottle from page 168, or make a new one. Have a partner quickly pump air into the bottle as you hold a hand on the bottle. What happens to the temperature of the bottle as air is pumped into it? What happens when the pressure is released?

_____

_____

16 What is the relationship between temperature and pressure in your bottle? (What happens to its temperature as you increase the pressure? What happens when you decrease the pressure?)

_____

_____

TIP:

Works best with a small bottle that is tightly sealed.

DISCUSSION:

17 Where is *pressure* usually highest and lowest in the atmosphere? Where is *temperature* usually highest and lowest in the atmosphere? What is the relationship between pressure and temperature as you move up or down in the atmosphere?

Why does squeezing air change its temperature?

You have to use energy to squeeze air or any other gas into a smaller space. Adding energy to a gas by squeezing it increases its temperature.

When a gas expands, it uses energy to push other stuff out of the way. When a gas loses energy, it cools down.

So, air warms up under high pressure as it is squeezed. Air cools down under low pressure as it spreads out.

This helps explain why air is usually warmer down low where pressure is highest, and colder up high where pressure is lower.

MORE TO TRY

Sometimes science demonstrations can seem like magic. Use what you've learned to help you figure out how each of the demonstrations below works.

18 Fill a bottle full of water and put the cap on tightly. Poke a hole in the bottle with a push pin. Water doesn't flow out of the hole unless you squeeze the bottle. What is holding the water in the bottle?

_____

_____

19 ★ If you remove the cap from the bottle in the problem above, water will flow out of the hole. Why does water flow out of the bottle when the cap is removed?

_____

_____

20 ★ If you've ever had a cup of cocoa with a lid like the one on the right, you may have noticed there is an extra hole in the lid. Why is there an extra hole in the cup lid?

_____

_____

21 When you use a can of spray paint, sunscreen, or any other liquid that has been mixed with gas and squeezed into a can, why does the can get cold?

_____

_____

22 Find a glass or hard plastic bottle with a wide opening. Fill the bottle to the rim and put a ping pong ball on top so that it is touching the rim. Flip the bottle upside down over a sink. If things go well, the ping pong ball won't fall off! (You may need to hold it at first.) What's keeping it there?

_____

_____

 DISCUSSION:

23 When you press a suction cup against a smooth surface, you create a small space with almost no air between the suction cup and the surface.

What is holding the suction cup against the surface?

Why does letting air into a suction cup pop it off?

Would a suction cup stick to the outside of a satellite in space?

WEATHER
CHAPTER 10:
Wind

For centuries, humans have harnessed the power of wind. Sailboats help humans travel around the globe. Windmills pump water, grind grains, and even generate electricity.

But why does the wind blow?

By the end of this chapter, you will know what causes the movement of air in the atmosphere.

=Lab   =Comic   =Enrichment

CHAPTER 11: HABITATS

A **habitat** is where a living thing makes its home. Sunlight, heat, water, and many other factors affect what lifeforms can survive and thrive in a habitat.

People from around the world marvel at the Grand Prismatic Spring in Wyoming, USA.

The water is heated to near boiling by an ancient underground volcano. You might think that nothing can survive in these hot pools, but life has found a way to survive!

What creatures might live in the unusual habitat pictured here?

CHAPTER 10:
WIND

Wind is the natural movement of air in the atmosphere.

Wind moves heat and moisture around the globe, causing many of the changes in weather we experience.

If you've ever spent the afternoon on the beach on a sunny day, you may have noticed that it often gets windy, and the wind almost always blows from the ocean towards the shore.

You can see this at the kite festival on the beach below. All of the kites are facing the wind that blows in from the sea.

At night, the wind usually switches direction, blowing out to sea.

We'll learn why in this chapter.

BOTTLED AIR

What happens to air when you heat or cool it?

To understand why the wind blows, we need to understand how air changes when it is heated and cooled.

MATERIALS

You will need:

- Empty plastic water bottle
- Two large, clear bowls, pitchers, or other deep containers that hold water
- Hot and cold water (ice if available)

1. Fill one bowl about halfway with hot tap water and the other with ice water. Leave enough room so that putting the bottle in will not cause the water to overflow.

2. Hold the empty plastic bottle in the hot water for at least 10 seconds with the top sticking out of the water. Screw the cap on the bottle while it is still in the water

3. Transfer the bottle to the cold water for at least 30 seconds. Watch and listen.

Switch the bottle back and forth between the hot and cold water and observe what happens.

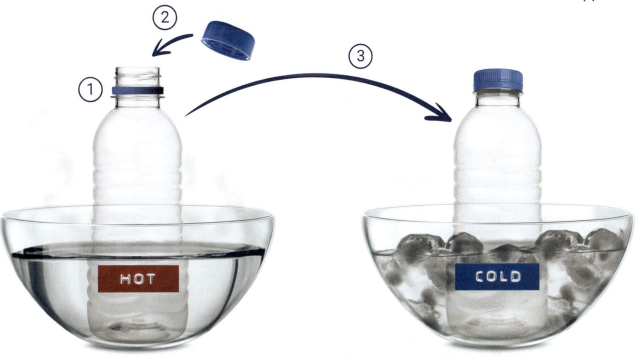

Answer the questions below about what happened.

1. After you place the cap on the bottle, can any air get in or out of the bottle? (Does the amount of air in the bottle change after the cap is on?)

_____

_____

2. What happens to the bottle when you move it from the hot water to the ice water and back again?

_____

3. How does the air in the bottle change when you move the bottle from the hot water to the ice water and back again?

_____

4. Which takes up more space: hot air or cold air? Explain your choice.

_____

_____

5. Do you think this experiment would work the same way with the cap off? Why or why not?

_____

_____

BALLOON BOTTLE

Try a few more demos to help you see the effects of heating and cooling air.

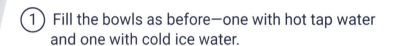

MATERIALS

You will need:

- All of the materials from the lab on page 178
- One balloon
- Thumb tack

① Fill the bowls as before—one with hot tap water and one with cold ice water.

② Stretch a balloon over the top of an uncapped empty bottle. It's best if the bottle is 500 mL (about 16 fl oz) or larger for these tests.

③ Make predictions for each question below before testing with your balloon bottle.

6 What do you think will happen to the balloon if you place the bottle into hot water? Why?

_____

_____

7 What do you think will happen to the balloon if you move the bottle from the hot water into the cold water? Why?

_____

8 Do both of the tests above. Describe what actually happens below.

_____

_____

Answer the questions below about other things you can do with your bottle, a balloon, and containers of hot and cold water. Predict what will happen, then try and see.

 9 Push an empty balloon inside your bottle and stretch the opening over the rim as shown. What will happen to the air in the bottle when you place it in cold water? What do you think will happen to the balloon?

_____

_____

 10 Poke a small hole in the cap with a thumb tack and screw the cap on to your bottle. What will happen to the air in the bottle when you place it upside-down in hot water? Will anything enter or leave the bottle?

_____

_____

11 Quickly switch the bottle in the previous problem from the hot to the cold water. What will happen to the air in the bottle when you place it upside-down in cold water? Will anything enter or leave the bottle?

_____

_____

 JOURNAL:

12 Come up with a variation of your own or find one online to try. Use different containers or new ways to heat the air inside, like a hair dryer. Describe your demo in your journal.

FULL OF HOT AIR

When air is heated, it spreads out. So, hot air takes up more space than the same amount of cold air.

Suppose you fill two balloons—one with hot air and the other with cold air. Since hot air takes up more space, it takes less hot air than cold air to fill the balloon to exactly the same size.

These hot air balloons in Turkey's Goreme National Park float in the air because the heated air inside the balloons is lighter than the surrounding air.

Balloons became the first vehicles to carry passengers above Earth's surface in 1783!

Since the balloon filled with hot air contains less air than one filled with cold air, the hot-air balloon weighs less than the cold-air balloon!

A hot air balloon rises in the atmosphere because the air inside is less dense, or 'lighter' than the surrounding air.

Answer the questions below about hot air.

13 The flame of a candle heats the air around it. Where is it easiest to feel this hot air?

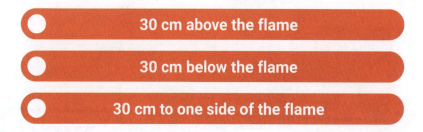

- 30 cm above the flame
- 30 cm below the flame
- 30 cm to one side of the flame

14 What direction does the hot air above a boiling pot on the stove usually go?

- Up
- Down
- Sideways

15 Which way does the hot, smoky air flow through a chimney?

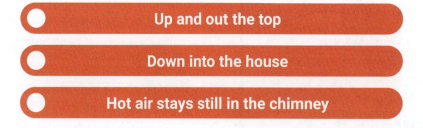

- Up and out the top
- Down into the house
- Hot air stays still in the chimney

16 Why is it best to place a smoke detector high on a wall or on the ceiling?

_____

_____

Air pressure is one reason air is colder up high and warmer down low.

But, there's another reason the air is usually warmer near the ground during the day.

Where do you think air in the atmosphere gets heated?

Probably way up high, where the air is closest to the Sun?

The Sun is so far away that being a few kilometers closer to it shouldn't matter much.

That's right, Winnie.

Besides, sunlight doesn't heat air very well.

THE SUN MAY LOOK LIKE IT'S RIGHT ABOVE THE CLOUDS, BUT IT'S ABOUT 150,000,000 KILOMETERS AWAY (SEE PAGES 130-131 FOR MORE). BEING 10 KM CLOSER TO THE SUN DOESN'T MAKE MUCH DIFFERENCE.

Since air is basically invisible, it doesn't absorb much sunlight.

Sunlight passes right through the atmosphere without heating the air much.

How does the air get warm, then?

Think of some things that **do** get hot in the Sun?

Name some.

185

187

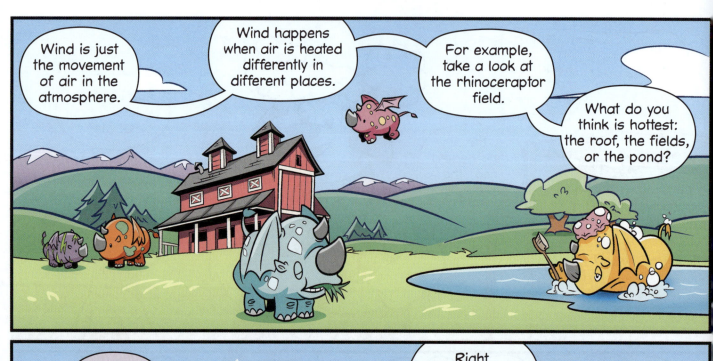

188

The mixing air creates wind.

When do you think breezes like these are usually strongest?

Probably on afternoons when the Sun is high and bright.

LOCAL BREEZES ARE OFTEN STRONGEST DURING THE DAY WHEN THE SUN IS HEATING THE SURFACE AND THE AIR IS MIXING. LARGE STORMS CAN CREATE WINDS THAT BLOW DAY OR NIGHT ACROSS HUGE AREAS.

That's why the rhinoceraptors are always flying in circles above the barn in the afternoon.

It's easier to fly in air that is rising!

That's right! Rhinoceraptors can circle for hours in rising columns of air called thermals.

Dragons, too!

RHINOCERAPTORS ARE NOT REAL, BUT ON EARTH YOU CAN SEE LARGE BIRDS CIRCLE IN RISING COLUMNS OF AIR TO SAVE ENERGY WHILE FLYING.

HEAT AND WIND

Much of the sunlight that reaches Earth passes right through the atmosphere and hits Earth's surface where it heats the land and oceans.

Sunlight doesn't heat the air much directly. Instead, air gets heated by land and water that have been warmed by the Sun. This is another reason that air is usually warmer down low than up high.

You can see the effects of heating on a sunny day when a hot surface heats the air just above it. The difference in temperature between the air at the surface and the air just above creates a mirror effect often called a mirage.

17 Where do you think it would be easiest to keep cool on a sunny day by the pool?

- ○ Lying on a towel placed directly on the concrete pool deck.
- ○ Sitting in a low chair a few inches above the concrete pool deck.
- ○ Sitting in a tall lifeguard chair four feet above the pool deck.

18 Which way does air flow if it is warmer and lighter than the air around it?

- ○ Down until it is the same temperature as the surrounding air.
- ○ Up until it is the same temperature as the surrounding air.
- ○ Hot air will not move unless the wind blows it.

Wind is caused by the uneven heating of Earth's surface by the Sun. Warm air rises in areas that are heated by the Sun. That warm air is replaced by cooler, heavier air that moves in to fill the space.

Wind can be a light backyard breeze or a huge current of air that circles the globe. The example below shows one way that uneven heating can create a breeze on a mountain that is heated by the Sun.

PRACTICE:

Use the diagram below to answer the questions that follow.

19 On a calm summer morning, the Sun has just begun to shine on the rocky face of the mountain above. Write the word WARM or COOL in the circled part of the mountain to describe whether the air there is getting warmer or cooler than the surrounding air.

20 Turn the three red lines above the circle into arrows to show which way the air above the circle will probably move.

21 Turn the three shorter red lines below the circle into arrows to show which way the air will probably move along the surface of the mountain.

DISCUSSION:

22 On a warm day when the mountain is covered in snow, the snow cools the nearby air. How do you think this might affect the flow of air near the circle above?

LAND AND WATER

On hot days, people flock to pools, lakes, rivers, and oceans to cool off in the water. Why does the water stay cool?

One reason is that water takes longer to heat and cool than most other materials. It also helps that water heated at the surface can mix with cooler water below. On land, the Sun only heats the surface. So, the temperature of water doesn't change as quickly as the land around it.

Sand and pavement can be cool at night, but too hot for bare feet during the day. The temperature of large bodies of water like lakes and oceans barely changes from day to night. So, we see a pattern along the coasts of oceans and lakes. During the day, bodies of water are usually cooler at the surface than the land around them. At night, the water is usually warmer than the land.

At Ipanema Beach in Rio de Janeiro, Brazil, water temperatures range from about 70°F (21°C) in August to 80°F (27°C) in January.

Air temperatures on the beach range from about 60°F (15°C) on the coolest nights to 95°F (35°C) on the hottest days. On hot days, the sand on the beach can reach more than 120°F (50°C).

The difference in surface temperatures between the water and the sand can create predictable wind patterns.

Answer the questions below by drawing arrows on each beach scene below.

23 On a summer afternoon, the Sun shines on the land and water above. Draw arrows to show where you think air will rise over hotter areas in the scene above.

24 Draw arrows to show where you think air will sink over cooler areas in the scene above.

25 Draw arrows near the shore line above to show which way you think the breeze will blow.

26 In many places, the surface of the ocean is warmer than the land at night. Draw arrows like the ones you drew on the daytime diagram above to show which way you think the air flows on the beach at night.

Answer the questions below.

27 Compared to other parts of the globe, do you think that air tends to rise or sink near the equator? Explain.

_____

_____

28 ⭐ Look at any globe and you'll notice that there is about twice as much land north of the equator as there is south of the equator. Do you think the northern hemisphere heats up and cools down faster or slower than the southern hemisphere? Explain.

_____

_____

29 ⭐ People who fly paragliders use winds that blow in from the ocean to soar above seaside cliffs in San Diego, USA. Do you think it's usually easiest to fly high above these cliffs in the morning or in the afternoon? Why?

_____

_____

30 Oceans absorb most of the sunlight that hits them. Almost no sunlight gets more than 200 meters below the surface. Do you think water is warmest near the top or the bottom of the ocean?

_____

_____

31 Flags and other objects let you know which way the wind is blowing. If you see flags on all four sides of a field blowing towards the field, what is probably happening to the air on the field?

_____

_____

_____

32 Towering storm clouds called thunderheads are the results of air rising in the atmosphere. Why are thunderstorms from these clouds most common during summer afternoons?

_____

_____

_____

UNIT 3:
SURVIVAL

CHAPTERS   PAGE

What does it mean to stay alive, or survive, in nature? In the next six chapters, we will explore how living things overcome the challenges they face to stay alive.

SURVIVAL
CHAPTER 11:
Habitats

ACTIVITY   PAGE

In the previous unit, we explored some of the reasons that different parts of the world have unique weather patterns, or climates.

How does the weather affect the foods and kinds of animals living in a location? Why do animals and plants from different regions look and behave differently?

By the end of this chapter, you will discover some of the reasons living things experience different challenges depending on where they live in the world.

AHA! = *Comic*   ➕ = *Enrichment*

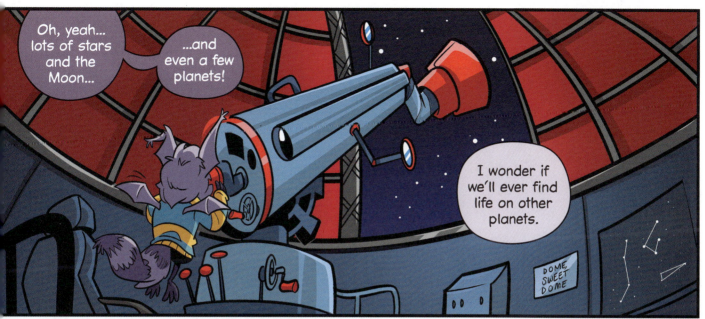

199

201

HABITATS

Below we compare several important qualities of six habitats.

Sunlight and **rainfall** affect the **heat** and availability of **food** in each habitat.

All of these qualities of a habitat affect the number of **predators**—animals that hunt other animals for food—that can survive there.

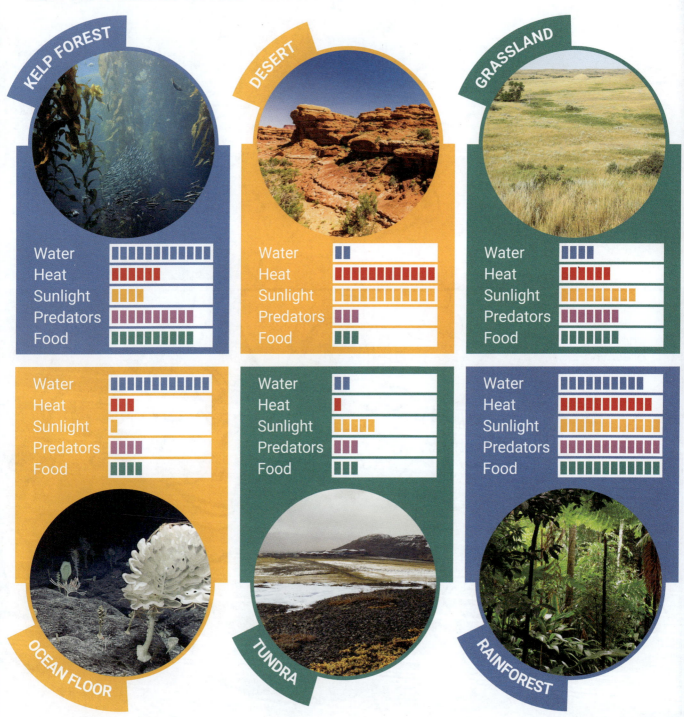

KELP FOREST

Water	████████████
Heat	█████
Sunlight	███
Predators	█████████
Food	████████

DESERT

Water	██
Heat	████████████
Sunlight	████████████
Predators	███
Food	███

GRASSLAND

Water	████
Heat	█████
Sunlight	█████████
Predators	███████
Food	██████

OCEAN FLOOR

Water	████████████
Heat	███
Sunlight	█
Predators	███
Food	███

TUNDRA

Water	██
Heat	█
Sunlight	████
Predators	██
Food	███

RAINFOREST

Water	████████████
Heat	████████████
Sunlight	████████████
Predators	████████████
Food	████████████

Use the information on the previous page to answer the following questions about habitats.

1 Which of these three habitats is driest?

⬤ Grassland ⬤ Tundra ⬤ Rainforest

2 Which of these three habitats is warmest?

⬤ Kelp forest ⬤ Ocean floor ⬤ Tundra

3 Plants need sunlight to grow, providing food for animals in a habitat. Do habitats with the most sunlight always have the most food? Explain.

_____

_____

4 What is a challenge some animals face even in habitats with plenty of food and water?

_____

_____

_____

DISCUSSION:

5 Which of these habitats has plenty of water but not much food? Why do you think there isn't much food in this habitat?

Creatures From The Deep

The farther you go below the ocean's surface, the darker it gets. Two hundred meters below the ocean's surface, there is barely any sunlight. And 1,000 meters below, there is no sunlight at all!

Researchers shine a light on this armored robinfish living at about 600 meters below the surface. Robinfish detect food in the dark with their feelers. Most of the nutrients here drift down from above.

FOOD WEBS

A healthy habitat has enough food for the living things there to eat. A **food web** is a diagram that shows what animals eat in their habitat.

PRACTICE:

Read the animal descriptions given for each food web to place each animal where it belongs in the web.

The arrows in a food web always point **from the food to the animal that eats it**. For example, pandas eat bamboo.

The arrows show how nutrients and energy flow from one living thing to another within a habitat.

PANDA

BAMBOO

6 Complete the tropical island food web using the animals described below.

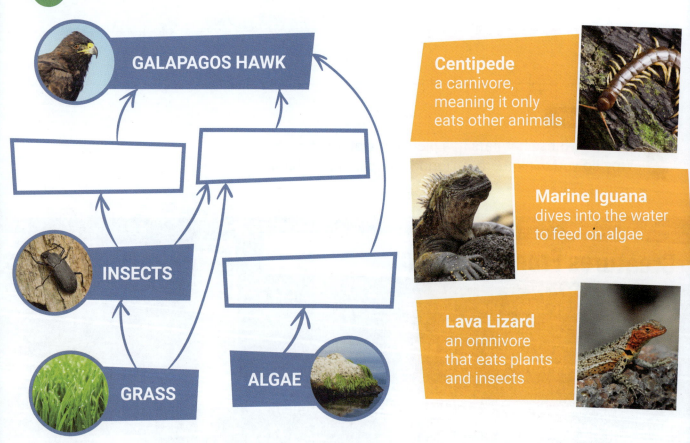

GALAPAGOS HAWK

INSECTS

GRASS

ALGAE

Centipede
a carnivore, meaning it only eats other animals

Marine Iguana
dives into the water to feed on algae

Lava Lizard
an omnivore that eats plants and insects

7 Complete the desert food web using the animals described below.

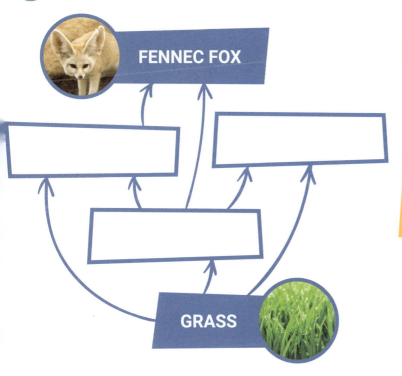

FENNEC FOX

GRASS

Ostrich
too large to be eaten by fennec foxes

Jerboa
an omnivore that eats plants, seeds, and insects

Locust
an herbivore, meaning it only eats plants

8 Complete the marsh food web using the animals described below.

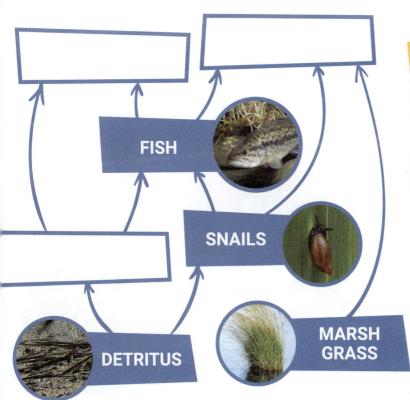

FISH

SNAILS

DETRITUS

MARSH GRASS

Spoonbill
an omnivore, meaning it eats both plants and animals

Crab
eats detritus, which is dead plant and animal material

Heron
a carnivore that eats small animals

9 Complete the tropical rainforest food web using the animals described below.

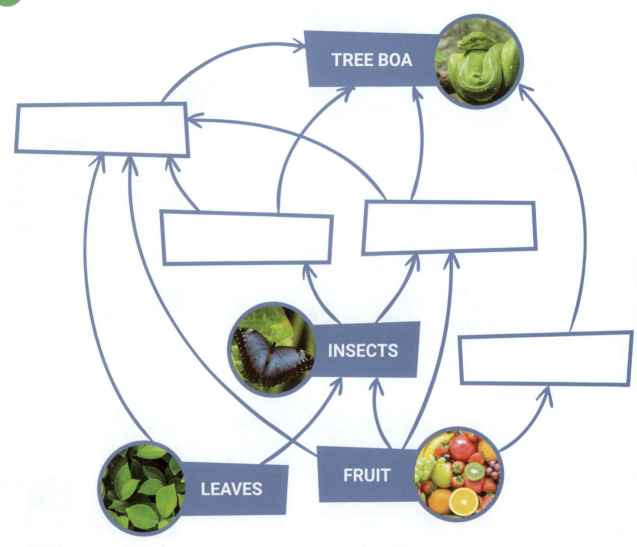

TREE BOA

INSECTS

LEAVES

FRUIT

Tree Frog
an insectivore, meaning it is a carnivore that mainly eats insects

Fruit Bat
a frugivore, meaning it is an herbivore that only eats fruit

Tanager
an omnivore that eats fruits and seeds as well as insects

Capuchin
an omnivore that eats a wide variety of plants and small animals like birds and frogs

10 Complete the coral reef food web using the animals described below.

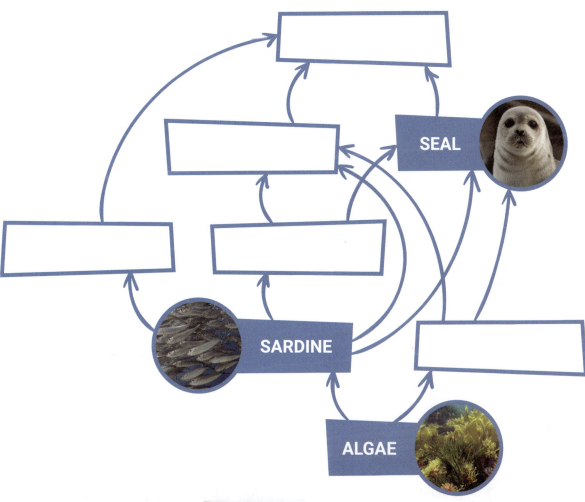

SEAL

SARDINE

ALGAE

Squid
a carnivore that catches and eats small fish

Tuna
a large carnivorous fish that eats a variety of smaller fish and squid

Anchovy
a small algae-eating fish that some people like to eat on their pizzas

Halibut
a flat fish that is so great at hiding, its only predators are sharks

Shark
a top predator that eats many different animals, but is not typically eaten by any other animals

CAUSE AND EFFECT

In the early 1900's, people hunted the wolves that lived in Yellowstone park. By 1926, there weren't any wolves left.

How did removing wolves change the numbers of plants and animals living in this habitat?

PRACTICE:

Use the diagram below to help you answer the questions that follow.

1895

WOLVES

BEAVERS

ELK

WILLOW TREES

GRASS

1995

ELK

BEAVERS

WILLOW TREES

GRASS

11 From 1895 to 1995, the number of elk ⬤ Increased ⬤ Decreased

12 From 1895 to 1995, the number of beavers ⬤ Increased ⬤ Decreased

13 From 1895 to 1995, the number of willows ⬤ Increased ⬤ Decreased

14 Why do you think the number of elk increased after wolves were removed?

_____

_____

15 Why do you think the number of willow trees decreased after wolves were removed?

_____

_____

16 If wolves returned, describe one way that the return of wolves might cause the number of beavers in the park to decrease.

_____

_____

17 In 1995, park rangers brought 31 wolves from Canada to Yellowstone park. After a few years, the number of beavers increased. Explain how the return of wolves could have caused the population of beavers to increase.

_____

_____

_____

RESEARCH:

18 Find out more by searching online for "wolves in Yellowstone Park" to see the most current information about this real-world project. The National Park Service (**nps.gov**) collects data about the animals and plants living in Yellowstone Park each year.

PYRAMID PUZZLES

We can think of the living things in a habitat as a pyramid.

At the base of the pyramid are living things that make food using energy from the Sun, like plants. In the middle are **herbivores**—animals that eat plants. At the top are **carnivores**—animals that eat other animals.

PRACTICE:

Fill every empty box with a letter that stands for a plant or animal, so that every animal has exactly the right amount to eat in the row below it. Order of letters within a row doesn't matter.

CARNIVORES (hawks, snakes, and foxes) go in the top row.
HERBIVORES (rabbits and mice) go in the middle row.
PLANTS (grass and berries) go in the bottom row.

HAWKS (H) eat 2 rabbits (R) or 3 mice (M)

SNAKES (S) eat 2 mice (M)

FOXES (F) eat 1 rabbit (R) or 2 mice (M)

RABBITS (R) eat 3 plots of grass (G)

MICE (M) eat 2 grass (G) or 1 berry (B)

In this example, the fox eats two mice. Each mouse eats two plots of grass.

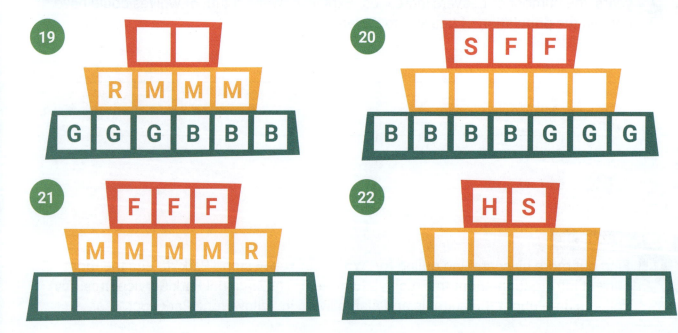

19

| R | M | M | M |

| G | G | G | B | B | B |

20

| S | F | F |

| B | B | B | B | G | G | G |

21

| F | F | F |

| M | M | M | M | R |

22

| H | S |

23

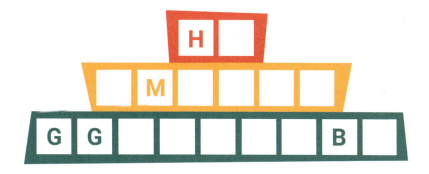

24

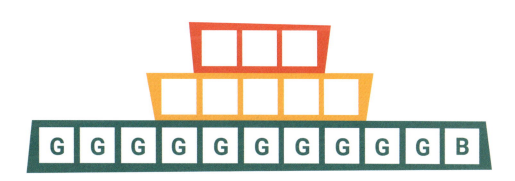

25 What is the greatest number of carnivores (H, S, or F) that can be in a pyramid puzzle with only 12 plots of grass (G)?

26 Why can't this puzzle be solved using the plants and animals given?

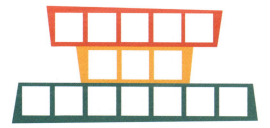

_____

_____

JOURNAL:

27 Make three Pyramid Puzzles and have a partner try to solve them. Use the animals given, or make your own rules for a different set of plants and animals.

DISCUSSION:

28 Do you think there are habitats with more carnivores than herbivores?

Which of the habitat types on page 204 do you think is most likely to have more meat-eaters than plant-eaters?

REVIEW

Every habitat has its own features and challenges. Use what you've learned about habitats to help you answer the following questions.

29 Describe how challenges in a habitat might change from summer to winter.

_____

_____

_____

30 Some animals only eat one kind of food. For example, pandas only eat bamboo. Explain why pandas might have more difficulty surviving than animals that eat lots of different foods.

_____

_____

_____

DISCUSSION:

31 Many animals sleep during the day and hunt for food at night. For example, owls hunt small prey mostly at night.

Do owls rely on the Sun for food? Why or why not?

What could possibly survive here?

There aren't any fish swimming in this water. It's way too hot! The center of the hot spring is almost boiling (around 90°C). The only living things you'll find in this hot spring are **microbes** (like bacteria). A single microbe is too small to see without a microscope, but billions of heat-loving microbes form the rainbow of colors seen in the hot springs above.

These microbes come in bright colors like red, orange, or yellow because they make carotene (carrot-teen), the same colorful pigment found in carrots and tomatoes.

Microbes can be found almost anywhere. They are able to survive in more habitats than any other form of life.

JOURNAL:

32 If we could view distant planets through a powerful telescope, would it be easier to see lifeforms that are similar to plants, animals, or microbes?

If we receive communication from distant aliens, do you think the aliens will be similar to plants, animals, or microbes?

If we find life on Mars, do you think it will be similar to plants, animals, or microbes?

Choose one of these scenarios and write a short story about discovering alien life.

SURVIVAL
CHAPTER 12:
Competition

When supplies are limited, the competition for survival heats up. What does it take to come out on top?

By the end of this chapter, you will understand the needs of living things and what it means to compete in nature.

🎲 = Game   AHA! = Comic   ✚ = Enrichment

CHAPTER 12: COMPETITION

Have you ever been in a competition? Maybe you have been in a race, an art contest, or the world championship of thumb wrestling?

In the wild, living things compete too. Plants and animals aren't competing for fame or prizes. They are competing to survive!

These hummingbirds swoop, stab, and dodge like flying swordfighters. Why are these tiny birds fighting with each other?

RESOURCES

Living things get the resources they need from their habitats.

Resources like water, carbon, oxygen, calcium, iron, and other materials help living things grow, develop, reproduce, and heal.

PLANTS

Trees, grasses, ferns, mosses, and shrubs belong to this group.

Plants need:
- Water
- Energy from sunlight
- Oxygen from water or air
- Carbon from air
- Minerals from soil (calcium, iron, and more)

ANIMALS

Animals need:
- Water
- Energy from food
- Oxygen from water or air
- Carbon from food
- Minerals from food (calcium, iron, and more)

Fish, birds, frogs, turtles, insects, snails, cats, dogs, and people belong to this group.

FUNGI

Molds, mushrooms, and baker's yeast belong to this group.

Fungi (fun-guy) need:
- Water
- Energy from food
- Oxygen from air (some don't need oxygen)
- Carbon from food
- Minerals from food (calcium, iron, and more)

Use the information on the previous page to help you answer each of the questions below.

1 What resources do animals get from their food?

_____

_____

2 This fish lives deep in the ocean where no sunlight reaches its habitat. Do you think plants grow there? Why or why not?

_____

_____

3 Growing plants without soil is called hydroponic gardening. What must be added to the water to keep these plants alive?

_____

_____

4 Mold is growing on an orange. Where does the mold get its energy?

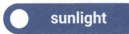 sunlight   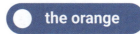 the orange    air

DISCUSSION:

5 Are fungi more similar to plants or animals in the ways they get resources?

221

Rhinoceraptors aren't the only territorial critters.

Elefinches sing songs to let other birds know, "Keep out! This tree is taken!"

Oh yeah! A pair made a nest in our yard last year.

They were mean!

Hippopotamooses fight over the best watering holes.

If you think an elefinch defending its nest is scary...

...you should see a hippopotamoose defending its muddy pond!

Some animals form groups that can protect a large area.

Like that pack of jackribbits that guards the cave up on the hill!

TERRITORY PUZZLES

Many animals have a territory—a part of their habitat that has resources they need. Territorial animals warn others to stay out of their space so they can avoid sharing resources like food, water, and shelter. In some cases, they will even chase away or fight animals that enter their territory.

PRACTICE:

In a Territory puzzle, place critters in a habitat by drawing dots in empty squares.

Wait! You can't just place them anywhere. Place each critter where it can reach food (🍎), shelter (⛰), and water (〰) with no sharing.

Critters can reach a resource that is next to the square they are in (diagonal is OK). Draw lines to connect each critter to its resources.

Ex. Place **2** critters in this habitat.

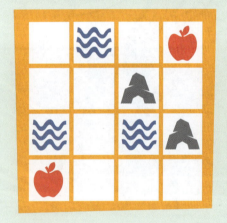 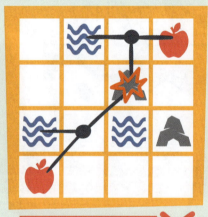 

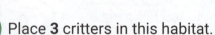
Nope. No sharing! ✗   Ah! That's better. ✓

6 Place **3** critters in this habitat.

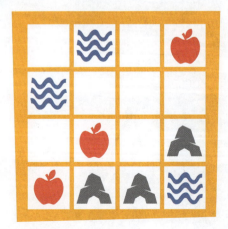

7 Place **3** critters in this habitat.

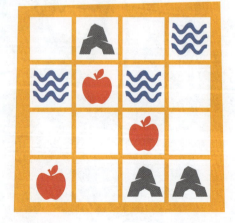

8 Place **3** critters in this habitat.

9 Place **3** critters in this habitat.

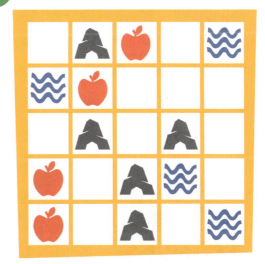

10 Place **3** critters in this habitat.

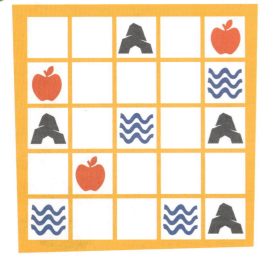

11 Place **4** critters in this habitat.

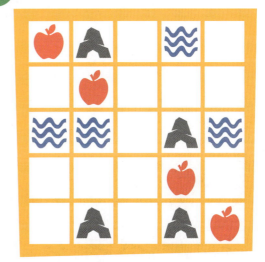

12 Place **4** critters in this habitat.

13 Place **4** critters in this habitat.

FIERCE COMPETITION

If two living things are both trying to use the same resource in a habitat, then we can say they are in **competition** for that resource. Competition can happen between members of the same species or between different species.

Jackals compete with other jackals, lions, and hyenas for food.

Gannets are territorial seabirds that compete for the best cliffside nesting spots. Laying eggs on the side of a cliff may not seem very safe, but here the eggs are out of reach for most predators.

Did you know that plants use their flowers to compete? Flowers with the brightest colors and best smells are most likely to be visited by hummingbirds and bees. Many flowering plants need help from animals to move pollen from flower to flower so they can make seeds.

 DISCUSSION:

14 Why do you think the hummingbirds on page 217 are fighting?

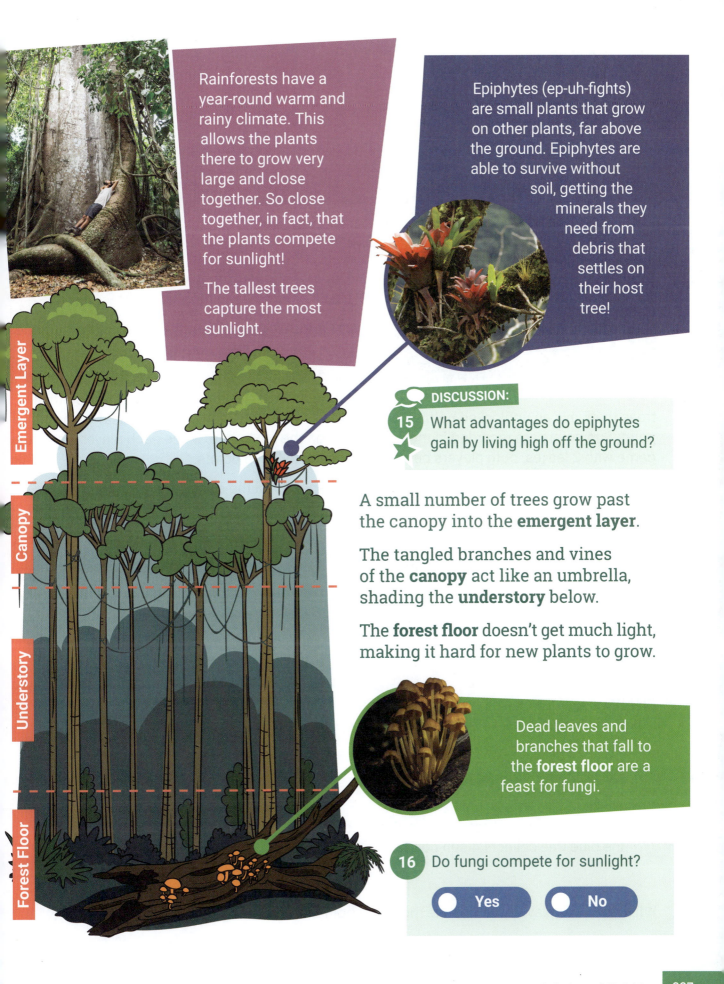

Rainforests have a year-round warm and rainy climate. This allows the plants there to grow very large and close together. So close together, in fact, that the plants compete for sunlight!

The tallest trees capture the most sunlight.

Epiphytes (ep-uh-fights) are small plants that grow on other plants, far above the ground. Epiphytes are able to survive without soil, getting the minerals they need from debris that settles on their host tree!

Emergent Layer

Canopy

Understory

Forest Floor

DISCUSSION:

15 What advantages do epiphytes gain by living high off the ground?

A small number of trees grow past the canopy into the **emergent layer**.

The tangled branches and vines of the **canopy** act like an umbrella, shading the **understory** below.

The **forest floor** doesn't get much light, making it hard for new plants to grow.

Dead leaves and branches that fall to the **forest floor** are a feast for fungi.

16 Do fungi compete for sunlight?

⚪ Yes   ⚪ No

GAME: CANOPY

Compete for sunlight in this game for two.

MATERIALS

You will need:

- Game Board (on next page)
- Pens or pencils in 2 colors

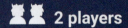 2 players

OBJECTIVE: Players take turns growing their trees to capture the most sunlight.

GAMEPLAY: Canopy is played in rounds using the game board on the next page.

SUN POINTS: To begin each round, players count the total number of leaves (filled circles) on their trees that have an unblocked path straight up to the sunlight. Each unblocked leaf earns 1 sun point. Sun points are recorded in the scoresheet on the next page. Since each player starts the game with 2 leaves, both players begin round one with 2 sun points.

GROWING TREES: In each round, Player 1 (blue) spends all the sun points they earned at the start of the round to grow their trees. Then, Player 2 (orange) grows their trees. Players can add leaves in two ways:

① **Growing straight up costs 1 sun point.**

Spend 1 point to connect any existing leaf of your tree to an empty circle directly above it.

② **Growing diagonally costs 2 sun points.**

Spend 2 points to connect any existing leaf of your tree to an empty circle above it diagonally left or right.

SHADE: A leaf that does not have a clear path to the top of the board is shaded. Shaded leaves do not earn sun points when they are counted at the beginning of each round.

Both of these trees have 8 leaves, but some of the leaves are shaded.

The orange tree collects 3 sun points and the blue tree collects 4 sun points.

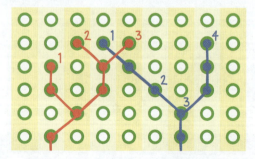

WINNING: The player who has the most sun points at the end of round 10 wins the game.

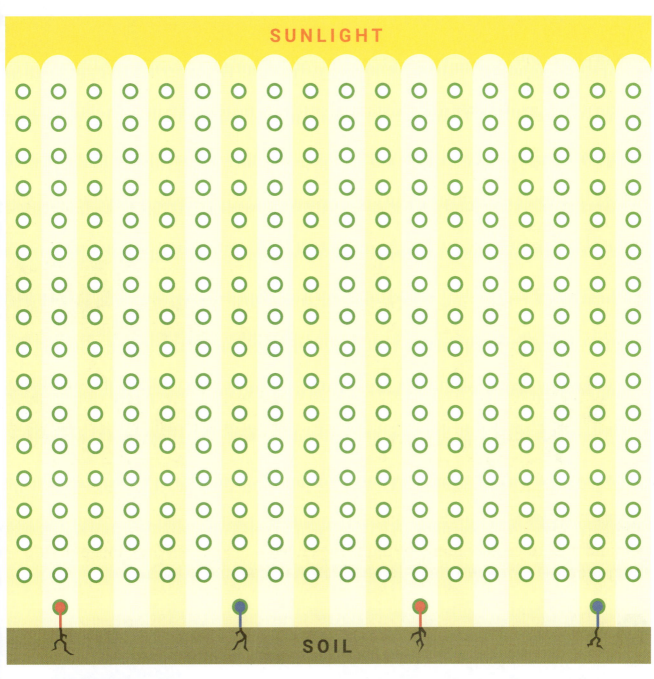

SUNLIGHT

SOIL

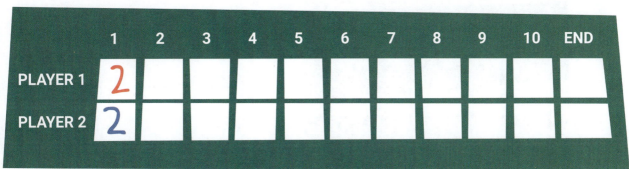

SCORECARD		1	2	3	4	5	6	7	8	9	10	END
	PLAYER 1	2										
	PLAYER 2	2										

Want a rematch? Print more Canopy game sheets at BeastAcademy.com

SUPPLY AND DEMAND

Supply is the amount of a resource like food or water that is available in a habitat. Supply of a resource in a habitat may change over time.

DISCUSSION:

17 The photos below were taken in the same park in New York City. Does the supply of plants available to eat change much from winter to summer? Does the supply of water change much from winter to summer?

Winter

Summer

Demand describes how many living things need a resource. When lots of living things compete for a resource, demand for that resource is high.

18 Flamingos get their pink color from pigments in the tiny shrimp and algae they eat. Circle the beach below that has a higher demand for shrimp and algae.

For each question below, write the word "high" or "low" to describe the supply and demand for the resource listed above each photo.

19 Trees for nests

Supply: [ ]   Demand: [ ]

Rainforests are home to thousands of species of plants and animals.

20 Water to drink

Supply: [ ]   Demand: [ ]

Herds of wildebeests and zebras travel across the dry savanna.

21 Fish to eat

Supply: [ ]   Demand: [ ]

This reef shark is surrounded by more fish than it can eat.

22 Fruit to eat

Supply: [ ]   Demand: [ ]

The tundra is too cold for most plants and animals to thrive.

DISCUSSION:

23 Which habitat above do you think has the most competition for the resource listed?

COMPETITION UPS AND DOWNS

Changes in supply or demand for a resource can affect competition in a habitat.

PRACTICE:

Use what you have learned about competition to answer the following questions.

24 Will competition for water be higher or lower than usual during a summer when precipitation is below normal? (Mark one.)

Competition will be... ◯ higher ◯ lower ◯ about the same

25 Fewer hummingbirds live in a local park this year than last year. Will competition for nest sites be higher or lower? (Mark one.)

Competition will be... ◯ higher ◯ lower ◯ about the same

26 During one month, there are fewer worms available for newts to eat. What will happen to the competition for worms if the number of newts stays the same? (Mark one.)

Competition will be... ◯ higher ◯ lower ◯ about the same

27 On a small island, lemurs eat insects and bats eat fruit. Mark all the changes below that could cause more competition for fruit.

◯ The bats there begin eating insects. ◯ The number of bats increases.

◯ The number of lemurs increases. ◯ The lemurs there begin eating fruit.

DISCUSSION:

28 If the amount of water in a habitat decreases, do you think there will be more or less competition for food? Explain.

Above is a habitat on Beast Island. This habitat is imaginary, but its food webs are similar to those on Earth. Here, chinchillipedes eat pine-apples. Porcupumas eat other animals, including chinchillipedes. Answer the questions below about this habitat.

29 A group of owligators moves into the habitat above. Owligators eat chinchillipedes. How could the owligators change the competition for food among porcupumas?

_____

_____

30 How could the owligators described above change the competition for food among chinchillipedes?

_____

_____

JOURNAL:

31 Describe how a new animal that enters the habitat above could decrease the competition for food among porcupumas? Can you think of more than one way?

SURVIVAL
CHAPTER 13:
Adaptations

The world is full of bizarre and beautiful plants and animals.

How do their parts and behaviors help living things survive? How can studying animal and plant adaptations help solve engineering problems?

By the end of this chapter, you should understand that the parts of each living thing have a purpose that helps them survive in a particular habitat.

🧪 = Lab   AHA! = Comic   ➕ = Enrichment

CHAPTER 13:
ADAPTATIONS

Every living thing has features that help it survive in the wild. Leopard spots, cactus needles, and cricket chirps are just a few. The features that help a living thing survive in its habitat are called **adaptations**.

A lethal hunter lies in wait for insects that come to feed from the flowers below. Can you find it?

With a bright pink body that looks like a flower, the orchid mantis easily blends into its habitat in the tropical forests of Thailand and Indonesia.

WELL ADAPTED

Living things face challenges like staying warm or cool, finding food, and avoiding predators. Adaptations are body parts and behaviors that help living things face these challenges.

PARTS

Penguins are birds that cannot fly. Instead, penguins swim and hunt in the water. Penguins move slowly on land, but that's okay because most of their predators like orcas and sharks hunt underwater. Let's take a closer look at how penguins' adaptations help them survive in their habitat.

PRACTICE:

For each penguin part below, choose whether its shape is more similar to another bird (a magpie) or an aquatic animal (a sea lion).

1 Penguin wings are more similar to ◯ **magpie wings**   ◯ **sea lion flippers**

 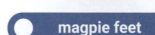 

2 Penguin feet are more similar to ◯ **magpie feet**   ◯ **sea lion feet**

3 Penguin bones are more similar to ◯ **magpie bones**   ◯ **sea lion bones**

BEHAVIORS

Some adaptations are **behaviors**—actions an animal chooses to do.

PRACTICE:

For each statement, underline the adaptation and mark whether it is a part or a behavior.

Ex. Fuzzy down feathers keep baby penguins warm on land.

 Part 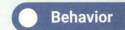 Behavior

---

4 Penguin chicks huddle together in groups to stay warm.

○ Part

○ Behavior

5 Slippery fish are easier to hold when you have a spiky tongue.

○ Part

○ Behavior

6 A fatty layer of blubber under their skin keeps penguins warm on land or in icy water.

○ Part

○ Behavior

7 If walking is too slow, penguins can slide on their bellies instead.

○ Part

○ Behavior

8 Penguins rub their feathers with oil (or preen) to stay waterproof.

○ Part

○ Behavior

9 Being waterproof is important because penguins spend a lot of time hunting for fish.

○ Part

○ Behavior

HANDS-ON ACTIVITY

What is it about the shape of our hands that makes them so useful?

MATERIALS

You will need:

- Tape
- Stopwatch (Learn how to use it on pages 60-61.)
- 2 tangerines or oranges

THE MIGHTY THUMB

Can you make an "O" shape by touching your fingers to your thumb? This task is impossible for most animals because most animals don't have thumbs that move like ours.

We share this unusual adaptation with monkeys, apes, koalas, and chameleons.

① Have your partner time how long it takes you to peel a tangerine or small orange.

[ ] seconds.

② With your partner's help, tape your thumbs to your hands so you cannot use them. (Not too tight!)

③ Have your partner time how long it takes you to peel a tangerine or small orange without using your thumbs.

[ ] seconds.

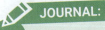
JOURNAL:

10 What other tasks would be hard without thumbs? With a partner, test how long it takes to perform at least three basic tasks with and without using your thumbs. Try writing a sentence, buttoning a shirt, or tying your shoes. Record the results in your journal.

LIMBS

Parts of an animal like arms, legs, fins, and wings that extend out from the body are called **limbs**. Limbs often have special shapes that help an animal move around its habitat.

Animals also use their limbs to find food, care for their young, and escape from danger.

PRACTICE:

Write the letter that matches each animal part to the task it is adapted to do best.

11 
mole

12 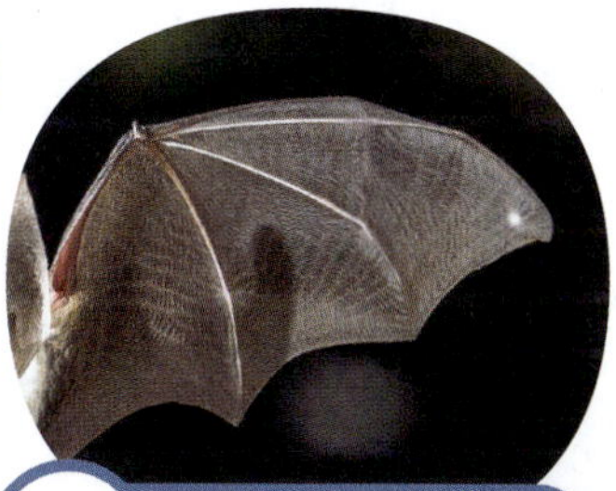
bat

A climbing

B flying

C digging

D swimming

13
river otter

14
chameleon

JOURNAL:

15 Choose one of the four tasks above or come up with one of your own.
Draw or describe how you could change the shape of your hands and arms to make them better adapted for the task. Be creative!

Solutions: 342-344

241

243

ANOLE LIZARD PARTS

Tiny **teeth** help anoles catch and eat small insects.

Green **scales** blend in with leaves to help anoles hide from predators.

Anoles signal to each other by bobbing their heads and showing off their bright red **dewlaps**.

Females lay about 6-18 **eggs** each summer.

Clawed fingers and toes grip the bark of trees.

TOMATO PLANT PARTS

Leaves use energy from the Sun to make sugars that the plant uses for food.

Bright **flowers** act as a signal to insects and birds that this tomato plant has something sweet to offer. These visiting animals help the plant pass its pollen from one flower to another.

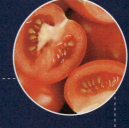

When pollen passes from one flower to another, the flower can form a fruit filled with seeds.

Some plants make harmful **toxins**.

Tomato leaves contain toxins that make them taste very bitter. Eating too many tomato leaves can make you feel sick.

Seeds become new baby plants.

Roots suck up water from the soil and hold the plant in place.

PARTS AND FUNCTIONS

Living things come in many shapes and sizes. Each part of a living thing has a function—a job to do. Some parts might have the same function, even though they look very different.

PRACTICE:

For each pair of plant and animal parts below, write what function the two parts share. Use the diagram on the previous page for clues.

Ex. What function do **plant seeds** and **animal eggs** share?

Making offspring (babies)

FUNCTIONS

- Getting food
- Taking in water
- Staying at the right temperature
- Moving around
- Communicating
- Making offspring (babies)
- Not getting eaten!

16 What function do **plant roots** and **animal mouths** share?

_____

17 What function do **plant leaves** and **animal mouths** share?

_____

18 What function do **plant toxins** and **animal scales** share?

_____

19 What function do **plant flowers** and **animal dewlaps** share?

_____

ADAPTATIONS MIX UP

Adaptations are useful, but only in the right situation. A fish out of water can't survive for long because it uses gills to breathe. Gills work by taking oxygen out of water, so gills work well in wet habitats but not in dry ones.

Each pair of adaptations in this activity has been mixed up. Can you fix them?

PRACTICE:

Circle the two blocks that must be switched to make both statements true.

Ex.

| Large ears help elephants | find their food | in the dark of night. |
| Large eyes help owls | keep cool | in the summer heat. |

(*The "Large ears help elephants" and "Large eyes help owls" blocks are circled.*)

20

| Lions use powerful fangs | to bite and hold | coral, rocks, or sand. |
| Octopus skin changes color | to blend in with | gazelles and zebras. |

21

| Rattlesnakes are | intelligent; | they can use tools and solve puzzles. |
| Crows are | venomous; | their fangs inject poison into their prey. |

22

| Hedgehogs have | thick fur to | stay warm in freezing weather. |
| Polar bears have | sharp spikes to | protect them from being eaten. |

23

| Snails have hard shells | to protect them from | their parents. |
| Baby birds make peeping sounds | to communicate with | predators. |

24

| Coconut seeds | spread to new places | with strong smells. |
| Flowering plants | attract insects like bees | by floating across the ocean. |

25

| Horses' hooves are useful for | grabbing branches | on wide open plains. |
| Monkeys use their flexible tails for | running quickly | in tropical forests. |

26

| Cacti store water in their stems | so they don't dry out | in the icy tundra. |
| Cushion plant leaves turn purple | to absorb extra heat | in the desert. |

27

| Sharks' teeth | shoot out of their mouths at high speeds | to catch bugs. |
| Chameleons' long tongues | fall out constantly | and are replaced by new ones. |

JOURNAL:

28 Choose two animal adaptations and write your own pair of mixed up sentences.

COLORS OF NATURE

Nature is rich with beautiful colors. But why? Can color be an adaptation?

TASTY COLORS

Plants are green because their stems and leaves contain green chlorophyll (klor-oh-fill). Chlorophyll uses sunlight to create sugars. Plants use these sugars for food—and so do animals that eat the plants.

SHOWY COLORS

Brightly colored flowers attract bees and butterflies that help transport pollen from flower to flower. This helps plants make seeds.

Brightly colored animals are able to attract the attention of mates in dense forests or murky water.

SNEAKY COLORS

Animals that blend into their environment are less likely to be seen and eaten by other animals.

Predators that blend in are also better at sneaking up on prey.

SCARY COLORS

Poisonous animals are often colored bright red or yellow. As a result, many animals avoid eating brightly colored animals like this poison dart frog.

Use the information on the previous page to help you answer the questions below.

29 A female bird with young spends a lot of time sitting in her nest. She must stay hidden from predators to survive. Use color clues to find and circle the female bird in each pair.

30 The bodies of these three insects have similar shapes and color patterns, but only one of them can sting. How could looking like a stinging insect help an animal survive?

_____

_____

_____

31 Venus fly traps are famous for catching flies with their "mouths" which are actually special leaves. How is color a clue these plants don't get all their food from flies?

_____

_____

_____

MASTERS OF DISGUISE

Blending in to your surroundings is called **camouflage** (cam-uh-flahj—rhymes with garage). Hiding is a useful survival strategy for many animals.

PRACTICE:

Can you match each animal to its photo?

A dead leaf butterfly

B ghost crab

C arctic fox

D thorn bug

E common octopus

F mossy frog

G ornate ghost pipefish

H copperhead snake

32

33

34

35

36

37

38

39

DISCUSSION:

40 How does looking like a flower help this orchid mantis catch insects?

BUILD A BEAST

Use your imagination and what you've learned about adaptations to build a beast and describe how it would survive in its habitat.

PLAY:

For each category below, roll a 6-sided die and mark the results for each roll by filling in the bubble that matches your roll.

ROLL 1. My beast moves using its...

| 1 | hooves | 2 | wings | 3 | tentacles |
| 4 | fins | 5 | claws | 6 | webbed feet |

ROLL 2. The number of limbs my beast has is...

| 1 | two | 2 | four | 3 | five |
| 4 | six | 5 | eight | 6 | ten |

ROLL 3. The color of my beast is...

| 1 | green | 2 | blue | 3 | gray |
| 4 | brown | 5 | white | 6 | orange |

ROLL 4. My beast's body is covered with...

| 1 | a hard shell | 2 | sleek scales | 3 | thick fur |
| 4 | smooth skin | 5 | glossy feathers | 6 | sharp spikes |

ROLL 5. My beast is known for its incredible...

| 1 | hooves | 2 | wings | 3 | tentacles |
| 4 | fins | 5 | claws | 6 | webbed feet |

Answer each question below about your cool creature.

JOURNAL:

41 Draw your beast in your journal. Use the results from the previous page.

42 Does your beast live on land, in the water, or both? How does it move around?

_____

_____

43 What does your beast eat? How does it get its food?

_____

_____

44 What does your beast protect itself from—predators, harsh weather, or something else? How does it protect itself?

_____

_____

45 Explain how the color of your beast helps it survive.

_____

_____

SURVIVAL
CHAPTER 14:
Feeding

Did you know teeth can tell you a lot about what animals like to eat?

One of the keys to survival is finding enough food to eat. How do animals' bodies help them eat their food?

By the end of this chapter, you will be able to recognize some interesting patterns between the shape of an animal's teeth and what it eats. But don't forget there are always exceptions to patterns in nature!

 = Comic   ✚ = Enrichment

CHAPTER 14: FEEDING

The shape of an animal's mouth and its teeth are specialized for the foods it eats.

Whale sharks are the largest living sharks on Earth. What adaptations do you think whale sharks have to help them get and eat their food?

This tooth belonged to a velocirabbit that lived long ago.

What do you think its teeth were used for?

That tooth is sharp, like a knife!

Velocirabbits must have eaten meat.

Great thinking! Even though there are no velocirabbits alive today...

...we think they ate meat because their teeth are similar to other meat-eating critters.

SIMILAR TO BEAST WORLD, SCIENTISTS ON EARTH COMPARE THE TEETH OF EXTINCT ANIMALS (LIKE DINOSAURS) WITH LIVING ONES TO GUESS WHAT ANCIENT ANIMALS MAY HAVE EATEN LONG AGO.

What kind of teeth are these?

TO BE CONTINUED...

TYPES OF TEETH

Teeth are adaptations that help animals eat the foods they need to survive. Animal teeth are similar to ours, but come in different shapes depending on which foods they eat.

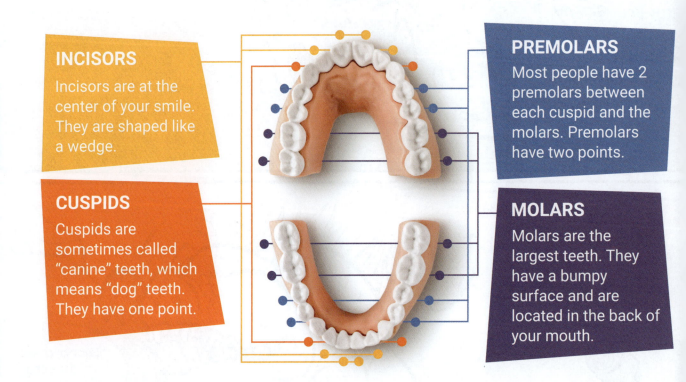

INCISORS
Incisors are at the center of your smile. They are shaped like a wedge.

CUSPIDS
Cuspids are sometimes called "canine" teeth, which means "dog" teeth. They have one point.

PREMOLARS
Most people have 2 premolars between each cuspid and the molars. Premolars have two points.

MOLARS
Molars are the largest teeth. They have a bumpy surface and are located in the back of your mouth.

PRACTICE:

Look at your own teeth in the mirror and feel the different surfaces each type of tooth has. Think about the ways you use your teeth when you're eating.

A

B

C

1 Your incisors are most like which of the tools above? _____

2 Your molars are most like which of the tools above? _____

3 Your cuspids are most like which of the tools above? _____

Use the pictures of these skulls to answer the questions on this page.

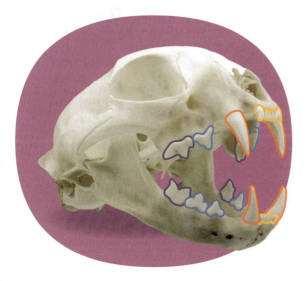

Bobcat   **Kangaroo**

4 Kangaroos have molars with small ridges for grinding plants, similar to our teeth. How are the shapes of a bobcat's molars different from a kangaroo's?

_____

_____

5 Bobcats have the same four types of teeth as us: molars, cuspids, premolars, and incisors. Which of these four types of teeth does this kangaroo NOT have?

_____

What's that huge gap?

Some animals have a large space between their incisors and molars, this gap is called a **diastema**.

A large <u>diastema</u> is more common in plant-eating animals.

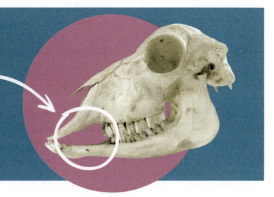

WHAT DO I EAT?

In this activity we'll look for evidence to help us predict what different animals eat. We can compare the teeth of animals to guess what foods they eat.

Evidence is information that helps us predict or explain something. Evidence might be an observation or a measurement.

PRACTICE:

Look closely at each animal's teeth. Place a check mark next to each adaptation that the top animal has. Then, circle the skull of the animal whose teeth (and probably diet) are most similar to the top animal.

Ex. Black bears eat both meat and plants. Circle the other animal that most likely eats both meat and plants.

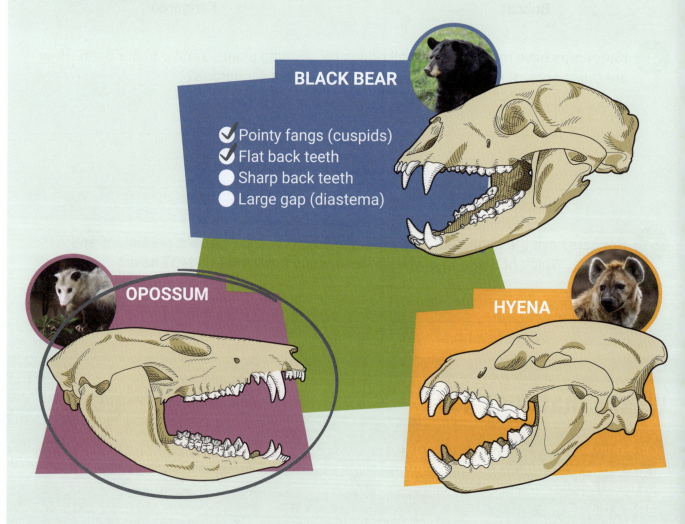

BLACK BEAR

- ☑ Pointy fangs (cuspids)
- ☑ Flat back teeth
- ◯ Sharp back teeth
- ◯ Large gap (diastema)

OPOSSUM

HYENA

6 Rhinoceroses eat plants. Circle the other animal that most likely eats plants

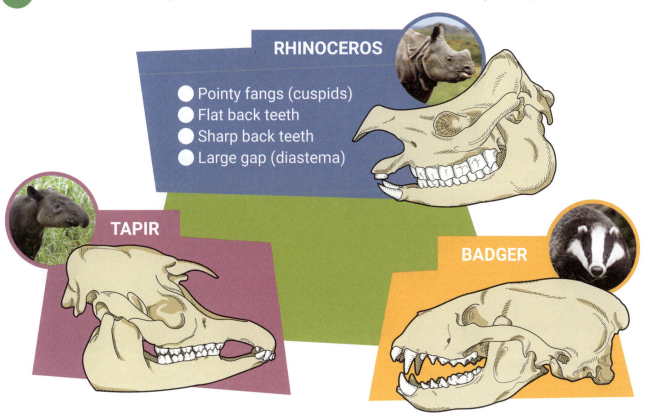

RHINOCEROS
- Pointy fangs (cuspids)
- Flat back teeth
- Sharp back teeth
- Large gap (diastema)

TAPIR

BADGER

7 Seals eat meat. Circle the other animal that most likely eats meat.

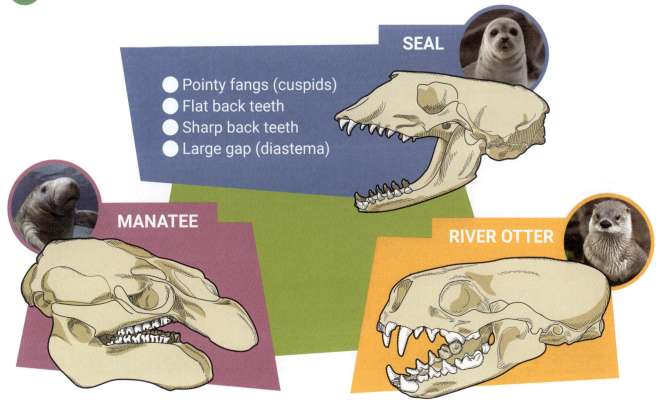

SEAL
- Pointy fangs (cuspids)
- Flat back teeth
- Sharp back teeth
- Large gap (diastema)

MANATEE

RIVER OTTER

 Solutions: 344-345

8 Hedgehogs eat insects. Circle the other animal that most likely eats insects.

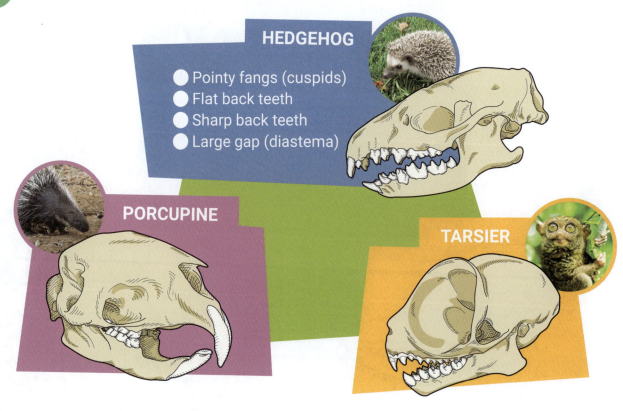

HEDGEHOG
- Pointy fangs (cuspids)
- Flat back teeth
- Sharp back teeth
- Large gap (diastema)

PORCUPINE

TARSIER

9 Rabbits eat plants. Circle the other animal that most likely eats plants.

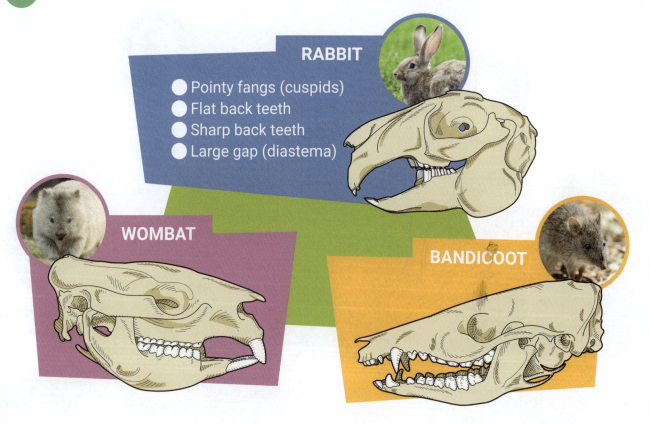

RABBIT
- Pointy fangs (cuspids)
- Flat back teeth
- Sharp back teeth
- Large gap (diastema)

WOMBAT

BANDICOOT

10 Giraffes eat plants. Circle the other animal that most likely eats plants.

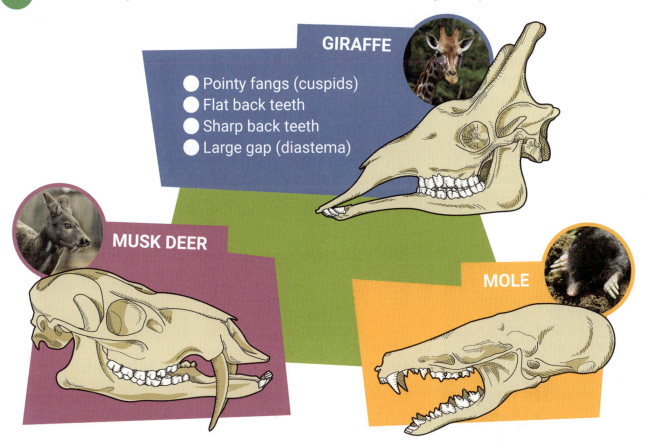

GIRAFFE

- Pointy fangs (cuspids)
- Flat back teeth
- Sharp back teeth
- Large gap (diastema)

MUSK DEER

MOLE

HANDY TEETH

Tusks are teeth (usually cuspids) that stick way out past the mouth. Walruses use their tusks to defend themselves, poke holes in ice, and pull themselves up like an ice climber.

TYPICAL TEETH

Looking at so many teeth, you might notice some patterns between tooth shape and what an animal eats. Scientists group animals into one of three groups based on what their bodies are adapted to eat.

PRACTICE:

Mark the check boxes below that match each animal's tooth adaptations. These are the types of teeth normally found in herbivores, omnivores, and carnivores.

HERBIVORE

(er-bih-vore or her-bih-vore)

Animals that eat plants.

OMNIVORE

(ahm-nih-vore)

Animals that eat meat and plants.

CARNIVORE

(car-nih-vore)

Animals that eat meat.

11
- ◯ Large fangs (cuspids)
- ◯ Flat back teeth
- ◯ Sharp back teeth
- ◯ Large gap (diastema)

12
- ◯ Large fangs (cuspids)
- ◯ Flat back teeth
- ◯ Sharp back teeth
- ◯ Large gap (diastema)

13
- ◯ Large fangs (cuspids)
- ◯ Flat back teeth
- ◯ Sharp back teeth
- ◯ Large gap (diastema)

TRICKY TEETH

Scientists change their explanations when they find more evidence.
Remember that each piece of evidence is only a clue.

PRACTICE:

Answer the questions below about this strange looking skull.

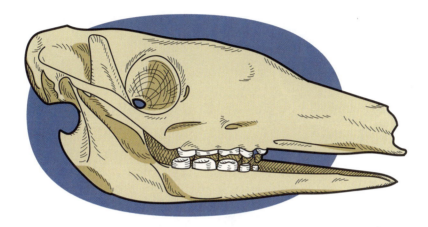

14 What types of teeth does this animal have?

⬤ Incisors   ⬤ Cuspids   ⬤ Molars

15 Based on its teeth, check all of the foods below that you think this animal might eat.

⬤ Fish   ⬤ Insects   ⬤ Fruit   ⬤ Grass

16 The skull above belongs to an aardvark. Aardvarks use their long sticky tongues to catch ants. Now that you have more evidence, what do you think an aardvark uses its teeth for?

_____

RESEARCH:

17 Find skull images of anteaters, pangolins, and echidnas. What do these skulls all have in common? How do you think these three animals eat?

NO TEETH, NO PROBLEM!

Most animals on Earth do not have teeth at all! Animals have adapted many different ways to capture and eat their food. These are just a few!

The box jelly below has caught a fish with its tentacles and pulled the fish into its body where it will be slowly digested.

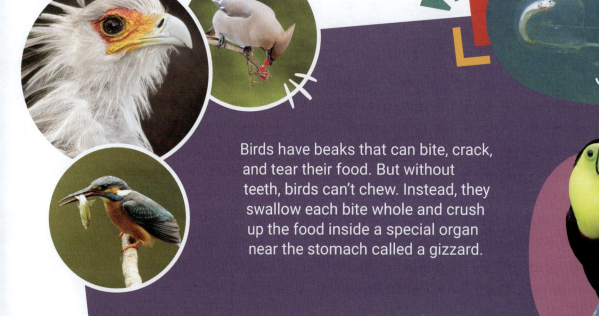

Birds have beaks that can bite, crack, and tear their food. But without teeth, birds can't chew. Instead, they swallow each bite whole and crush up the food inside a special organ near the stomach called a gizzard.

18 Insects do not have teeth, but they have all kinds of interesting mouths.

Match each insect mouth to the tool that it is most similar to.

A

B

C

Feather-duster worms have feathery tentacles around their mouths that catch tiny plants and animals drifting in the ocean called plankton. Many plankton are so small you need a microscope to see them!

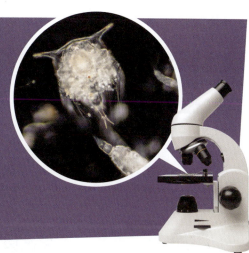

A **filter** works by catching some materials and letting others pass through tiny spaces or holes in the filter. **Filter feeders** are animals that use part of their bodies to filter food from the water.

19 Which of these tools is a filter? _____

A B C

Many of the largest animals on Earth are filter feeders that dine on the ocean's tiniest animals.

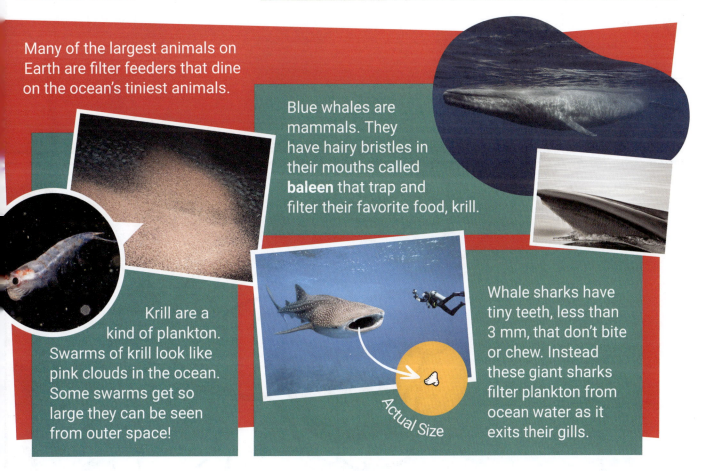

Blue whales are mammals. They have hairy bristles in their mouths called **baleen** that trap and filter their favorite food, krill.

Krill are a kind of plankton. Swarms of krill look like pink clouds in the ocean. Some swarms get so large they can be seen from outer space!

Actual Size

Whale sharks have tiny teeth, less than 3 mm, that don't bite or chew. Instead these giant sharks filter plankton from ocean water as it exits their gills.

Ms.Y. TEETH PART 2

Did you ever catch a leopard seal on camera?

Yes!

It took 3 weeks but we finally spotted one.

Let's watch the video.

There's so many krill it looks like a pink cloud!

There's the leopard seal!

AM 09:17

It sucked up the krill and spit the water out through its teeth.

This was the first ever evidence that leopard seals can filter feed!

Are you for real?

The leopard seals in Beast World are imaginary, but there are real animals called leopard seals living on Earth.

These are photos of leopard seals in Antarctica. Like Ms. Y., scientists recently discovered that leopard seals use their teeth to not only catch fish and penguins, but also to filter tiny krill from the water.

SURVIVAL
CHAPTER 15:
Protection

What dangers lurk in each habitat and how can the creatures living there avoid these dangers?

By the end of this chapter, you will be familiar with a few strategies that animals and plants use for defense against harsh weather and hungry predators.

🧪 = Lab   🎲 = Game   ➕ = Enrichment

CHAPTER 15:
PROTECTION

Each habitat has unique challenges, such as extreme temperatures or fierce predators. In this section, we'll explore how plants and animals protect themselves from the dangers in their habitat.

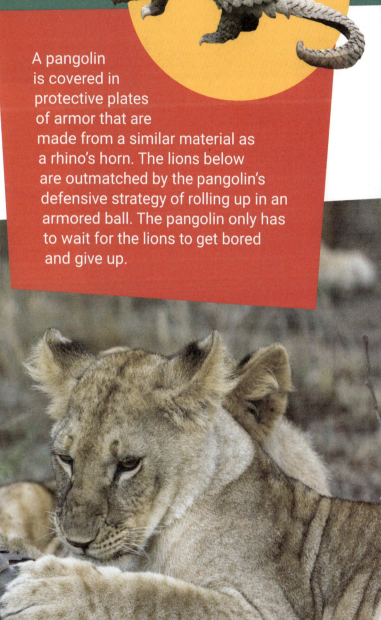

A pangolin is covered in protective plates of armor that are made from a similar material as a rhino's horn. The lions below are outmatched by the pangolin's defensive strategy of rolling up in an armored ball. The pangolin only has to wait for the lions to get bored and give up.

CHICKEN RUN

Do you think you could outrun a chicken? How about a guinea pig? Compare your running speed with the speed of other animals.

MATERIALS

You will need:

*   Stopwatch (Learn how to use it on pages 60-61.)
*   50-foot path (the width of a basketball court)

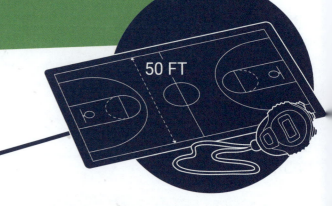

50 FT

ON YOUR MARK

1. Mark a path that is 50 feet long. We recommend using the width of a basketball court.

2. Use a stopwatch to measure how many seconds it takes to run 50 feet.

3. Repeat the test 3 times and record your results on the next page.

4. Use the Speed Table on the next page to find your fastest speed in miles per hour (mph).

How long did it take to run 50 ft?

1st try _____ seconds

2nd try _____ seconds

3rd try _____ seconds

Time to run:

50 FEET	SPEED	ANIMAL
1 second	34 mph	Fox
2 seconds	18 mph	Goat
3 seconds	12 mph	Squirrel
4 seconds	9 mph	Chicken
5 seconds	7 mph	Mouse
6 seconds	6 mph	Penguin
7 seconds	5 mph	Guinea pig
8-9 seconds	4 mph	Hedgehog
10-13 seconds	3 mph	Hamster
14-18 seconds	2 mph	Turtle

1 Use the table on the right to figure out about how fast you can run in miles per hour (mph)?

NEED FOR SPEED

The warthog below is running from a hungry lion. Speed is one adaptation animals use to defend themselves from predators.

TIRED OUT

After only a few minutes of running, this lion lays down in the shade and pants to cool down her body. A lion's top speed is about 50 mph (81 km/h), but it can only keep this speed for a few seconds at a time.

✏ **JOURNAL:**

2 Other than running faster, what other adaptations might animals use to defend themselves from a lion. List as many as you can think of in your science journal.

GAME: PREDATOR & PREY

No animal wants to be someone's lunch. Many adaptations help animals avoid being captured and eaten by other animals.

MATERIALS

You will need:

- One coin
- One deck of prey cards, and one deck of predator cards*

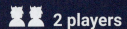 2 players

GAMEPLAY: One player will play as the Predator and the other player will play as the Prey. Each game is played in nine rounds called "hunts."

Each player will shuffle their own deck of cards. To begin, each player draws a card from their deck without showing the other player. Each player then announces only the name of their animal.

In the first hunt, the Prey player goes first and announces whether they choose to hide, fight, or flee. Both players reveal their cards. Players compare the chosen ability for the hunt as below. The higher ability wins.

HIDE Compare each animal's ability to hide or detect.

FIGHT Compare each animal's ability to fight.

FLEE Compare each animal's ability to flee or chase.

If Predator and Prey's abilities are tied, flip a coin. If heads, Prey wins; if tails, Predator wins.

Each player draws a new card from their deck. This time, the Predator announces whether they will detect, fight, or chase. Players reveal their cards to see who has won the hunt.

Play continues with players taking turns choosing which ability to compare.

WINNING: The player who wins the most matches wins the game.

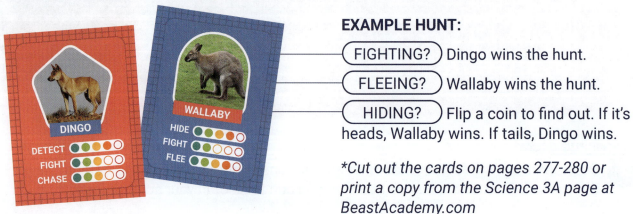

EXAMPLE HUNT:

FIGHTING? Dingo wins the hunt.

FLEEING? Wallaby wins the hunt.

HIDING? Flip a coin to find out. If it's heads, Wallaby wins. If tails, Dingo wins.

*Cut out the cards on pages 277–280 or print a copy from the Science 3A page at BeastAcademy.com

FOX

DETECT
FIGHT
CHASE

COUGAR

DETECT
FIGHT
CHASE

WOLF

DETECT
FIGHT
CHASE

BOBCAT

DETECT
FIGHT
CHASE

EAGLE

DETECT
FIGHT
CHASE

WEASEL

DETECT
FIGHT
CHASE

COYOTE

DETECT
FIGHT
CHASE

FALCON

DETECT
FIGHT
CHASE

SNAKE

DETECT
FIGHT
CHASE

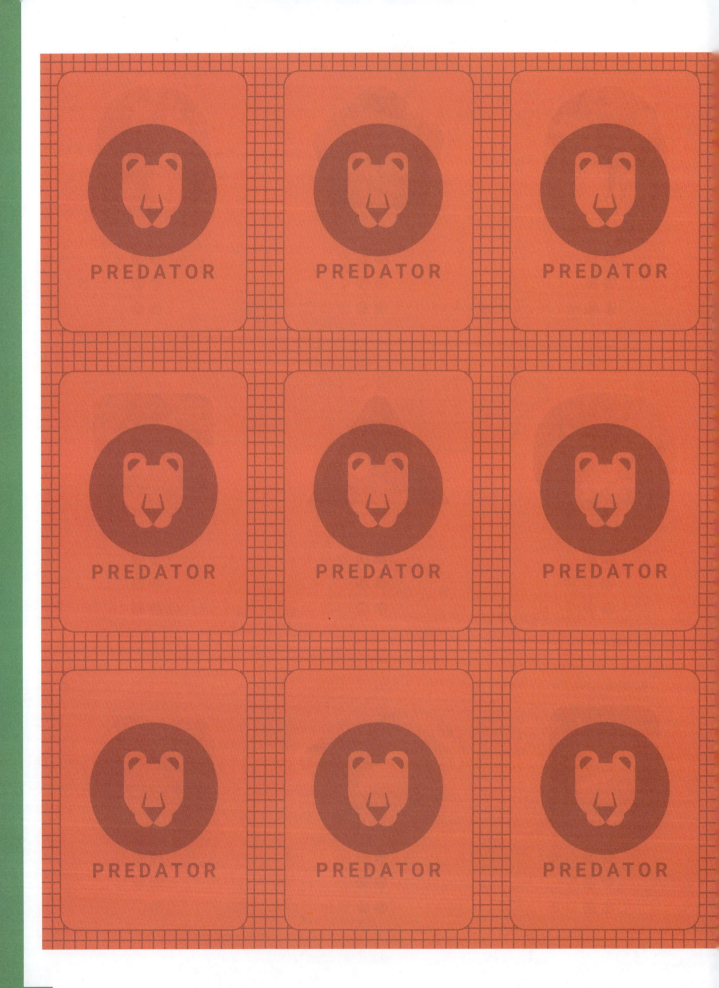

VARIATIONS

In the game of Predator and Prey, the fox card has circles that show its abilities. In reality, foxes aren't all exactly the same. One fox may have a better sense of smell or longer legs than the others. These small differences in the bodies or abilities of foxes (and all other types of living thing) are called **variations**.

Which variations can help living things survive?

3 Moths are active at night and often spend the day resting on trees. There are three moths in this image.

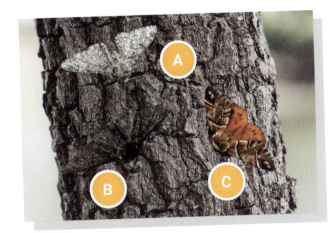

Which moth is most likely to be eaten by a bird?

Which moth is least likely to be eaten by a bird?

4 Explain why you chose the moths in the previous questions.

_____

_____

5 Which of the three moths would have the best chance of surviving while resting on a birch tree like the one pictured here?

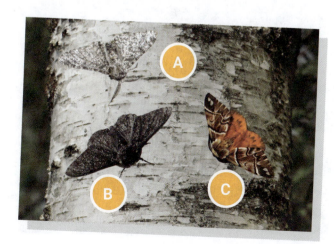

Body size can be a variation. In the photo below, we see fish that come in a variety of sizes. In nature, do you think it's more useful to be large or small?

6 Explain how the size of fish A could help it avoid getting eaten by the barracuda.

_____

_____

_____

7 Explain how the size of fish B could help it avoid getting eaten by the barracuda.

_____

_____

_____

Answer the questions below about the role of thorns on berry bushes.

8 Animals that eat berries often pass live seeds out in their droppings, allowing plants to spread to new places. Which of the plants above is *most likely* to have its berries eaten and spread by lots of animals?

9 Deer avoid large thorns, but will carefully eat berries from a plant with small thorns. They will eat the entire plant if there are no thorns. Which plant above has the *best* protection against being eaten by deer?

10 Mice eat berries and also hide in plants with thorns of any size to escape cats and foxes. Which plant above would give mice the *best* protection from predators?

11 Match each berry bush above to the location below where the bush would most likely be found. Write A, B, and C once each in the circles below.

A park with only deer

A park with only mice

A park with mice and foxes

DEFENSE

Let's take a look at some defensive adaptations used by different animals, plants, and fungi.

DON'T EAT ME!

Poison is a chemical weapon that some living things use to defend themselves. Many kinds of mushrooms, like these fly agarics, are poisonous. Boxer crabs carry a stinging anemone in each claw. Their tiny toxic "boxing gloves" protect them from hungry fish.

Ouch! Sharp spikes make a hungry animal think twice about taking a bite of this porcupinefish. Cactus needles protect cacti from being eaten. Water drops on the needles tend to slide towards the body of the cactus, helping it collect water in very dry climates.

The mimosa plant closes its leaves and droops down when touched. Small, drooping leaves don't look as tasty to hungry herbivores. Some animals will pretend to be dead to avoid being eaten. This hognose snake is very dramatic, don't you think?

BEAT THE HEAT!

In very hot habitats with little rain, animals must defend themselves from drying out. The smooth scales of lizards and snakes block water from escaping their bodies.

Many desert animals, like this jerboa, spend time in underground burrows during the day when the sun is brightest and predators like snakes are active.

WHITE HOT

Near the poles and high in the mountains, we find animals battling extreme cold!

You've probably heard it's best to wear light colored clothing in summer because it stays cool. Why then are most arctic animals white?

You might guess that white fur is good camouflage in the snow, and you would be right! But white fur can also be useful for staying warm. A polar bear's white fur is actually **transparent** (or see-through). Sunlight that hits the fur gets scattered in random directions, keeping a lot of heat trapped near the skin as it bounces between hairs.

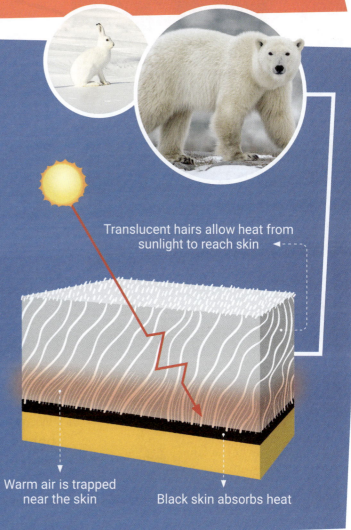

Translucent hairs allow heat from sunlight to reach skin ◄

Warm air is trapped near the skin

Black skin absorbs heat

DISCUSSION:

12 What are some examples of adaptations that protect a living thing from more than one danger in its habitat?

INSULATION LAB

You've probably never stopped to think about how a jacket keeps you warm when it's cold outside. Jackets work as insulation. Let's explore how insulation works.

MATERIALS

You will need:

- Pitcher of warm water between 40-50°C (100-120°F)
- 3 small water bottles
- Thermometer
- Bubble wrap
- Rubber bands
- Timer or watch
- Cold place, like a fridge

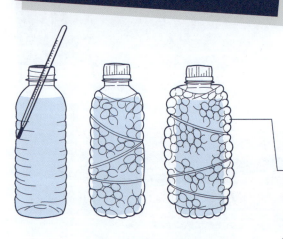

1. Cut out three identical sheets of bubble wrap, each large enough to cover the bottom and sides of a water bottle.

2. Carefully fill all three water bottles with hot water. **Water temperature must be less than 50°C (120°F) to prevent burn injuries!**

3. Wrap one water bottle with one sheet of bubble wrap, covering the bottom and sides. Use rubber bands to hold it in place.

4. Wrap a second bottle with bubble wrap as before, but this time use two sheets of bubble wrap to cover the bottom and sides.

5. Measure the temperature of all three bottles and place them somewhere cold and dry. A refrigerator or freezer works great.

6. Measure the temperature of all three bottles every 10 minutes and record it on the next page. Replace the cap after each measurement.

13   Record your data in the table below.

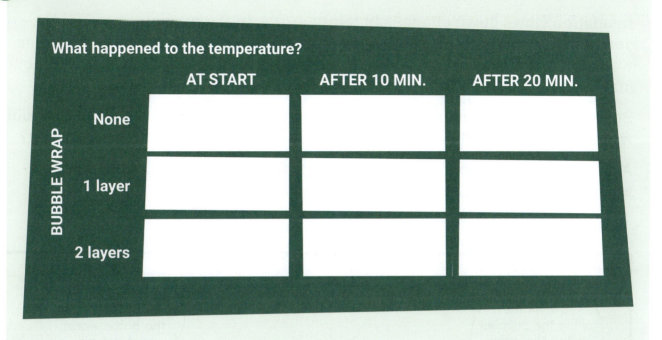

What happened to the temperature?

BUBBLE WRAP	AT START	AFTER 10 MIN.	AFTER 20 MIN.
None			
1 layer			
2 layers			

14   How much did the temperature *change* after 20 minutes in each water bottle? Write your answers in degrees Celsius (°C) or degrees Fahrenheit (°F).

No bubble wrap	1 layer of bubble wrap	2 layers of bubble wrap

15   Which bottle had the smallest **change** in temperature after 20 minutes?

_____

16   Increasing the amount of bubble wrap changed how fast the water cooled down. Which part of the bubble wrap do you think causes this to happen?

 The plastic   ○ **The air inside the bubbles**

HOW DOES IT WORK?

Insulation slows down changes in temperature. In the previous experiment, you probably found that a second layer of insulation helped to keep the water warmer. This makes sense. You've probably put on a second layer of clothing to stay warm.

But why does a second layer work better than just one? Does the air in the bubble wrap matter, or would bubble wrap keep a bottle warm even if it was popped?

Alex and Lizzie set up a new experiment to test popped vs unpopped bubble wrap to see whether the air in the bubble wrap makes a difference.

17 Predict what you think will happen if a warm bottle of water with popped bubble wrap and a warm bottle of water with unpopped bubble wrap are placed in the cold. Will one cool down faster than the other? Write your prediction in the space below.

18 Repeat the lab on pages 286-287, but this time use two layers of popped and two layers of unpopped bubble wrap. Record your data in the table below.

What happened to the temperature?

BUBBLE WRAP	AT START	AFTER 10 MIN.	AFTER 20 MIN.
None			
2 layers popped			
2 layers unpopped			

19 How much did the temperature change in each water bottle? Write your answers in degrees Celsius (°C) or degrees Fahrenheit (°F).

No bubble wrap

2 layers of popped bubble wrap

2 layers of unpopped bubble wrap

20 Based on the results of your experiment, explain how bubble wrap works to keep a water bottle warm.

_____

_____

_____

REVIEW

Your body makes its own heat constantly, warming the air just above your skin. But that warm air can simply blow away, leaving cold air in its place. Fuzzy down feathers and fur both do a great job of trapping air warmed by body heat and keeping it in place.

People have been making clothing from sheep's wool for nearly 5,000 years!

Down feathers are the first feathers to grow. That's why baby birds are so fuzzy. Jackets filled with down feathers can keep you very warm.

21 When you are wearing a jacket, where does the warmth in the jacket come from?

- Down feathers
- Fabric
- Your body

22 Explain how bubble wrap in the insulation lab works like animal fur or feathers.

_____

_____

_____

23 Here are three insulation materials that block heat from moving from one place to another. What do they have in common that makes them good insulators?

STYROFOAM

FIBERGLASS

PACKED SNOW

- ◯ All three are made of the same material.

- ◯ All three materials have tiny spaces filled with air.

- ◯ All three materials are able to create their own heat.

24 A styrofoam cup keeps hot foods hot by preventing heat from escaping. Styrofoam can also keep cold foods cold. How?

_____

_____

_____

DISCUSSION:

25 Reptiles, like lizards and snakes, are often called "cold-blooded" because they don't generate enough heat to keep their bodies warm like birds and mammals do. Instead, reptiles sit in the sun to warm their bodies.

Do you think a sweater could help keep a pet lizard warm on a cold day? Explain.

SURVIVAL
CHAPTER 16:
Cooperation

How do different kinds of living things work together to survive?

By the end of this chapter, you should realize that many challenges, whether they are in the classroom or in nature, can be easier to face when you are part of a group.

🎲 = Game AHA! = Comic ➕ = Enrichment

CHAPTER 16:
COOPERATION

So far we've seen that nature can be a challenging place to survive. One of the best strategies for staying alive is working together. Some animals form large groups, such as herds of elephants or schools of fish. Other animals form more unusual relationships.

The world's largest rodent, the capybara, is native to the forests and savannas of South America. Capybaras spend much of their time in the water.

In the wild, you might spot a bird riding atop a capybara. The capybara doesn't seem to mind. What's going on here?

STRENGTH IN NUMBERS

Many animals form groups that help them catch more food than they could alone, defend themselves from predators, and stay warm or cool in harsh weather.

A group of dolphins (called a pod) surrounds a school of fish, making it easier to hunt than if the dolphins were alone.

Outside of their underground tunnels, some meerkats will look out for predators and let out a squeak to warn their whole group (called a mob).

A group of partridges (called a covey) huddles together to stay warm. Huddling works because part of each animal's body is guarded from the cold air. What a cozy covey!

 RESEARCH:

1 Use the internet to find out which of the following animals form groups. If so, what is their group called? **monkeys, ravens, koalas, geese, narwhals, hornets**

Search for "group of monkeys is called" or "collective noun for monkeys."

What animals form the largest groups?

The largest groups of animals in the world are colonies of ants. A single colony can include millions of ants! Within a colony, each member has a job to do. The bodies of different ants are specialized for the kinds of jobs they have.

Queen ants are large ants born with wings. When she starts her own nest, the queen loses her wings and begins laying eggs. Some queen ants can lay as many as one million eggs!

Worker ants are small female ants that can't lay eggs of their own. Instead, they care for the queen ant's babies—the worm-like larvae pictured here. Worker ants also build tunnels and forage for food.

Male ants have wings and fertilize the eggs laid by a queen ant.

Soldier ants can be much larger than worker ants. These huge female ants defend the nest from predators and can carry large pieces of food back to the nest.

GAME: FORAGE

Players compete to gather as much food as they can to store in their nests.

MATERIALS 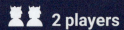 2 players

You will need:

- Two dice
- Two sets of 6 tokens (we suggest 6 red and 6 black checkers)
- 30-36 dry beans or corn kernels
- Game board (following page)

OBJECTIVE: You control six ants living in a colony. Collect as many beans as possible and return them to your nest. The colony that gathers the most beans wins.

SETUP: Each player places a stack of 6 ant tokens (checkers) on their Nest space.

For each food space on the board, roll 2 dice and place that many beans on the space.

GAMEPLAY: Players take turns moving any of their own tokens a total of 6 spaces.

Here are three different ways a player might move 6 spaces:

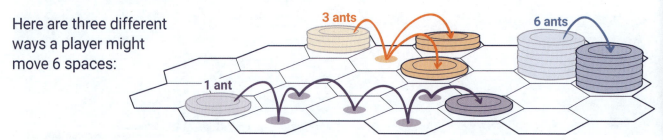

COLLECTING FOOD: When a stack of ant tokens lands on a space with beans, they can collect them. Beans are carried according to the rules below, including when stacks split or merge. Beans can be dropped off at the nest for safe storage.

Maximum beans a group can carry:	1 bean	2 beans	4 beans	5 beans	7 beans	9 beans
	1 ant	2 ants	3 ants	4 ants	5 ants	6 ants

ATTACKING: A stack of ant tokens can attack any opponent's stack that is the same size or smaller by landing on it. The attacked ants must drop any beans they are carrying onto the space and return to their nest empty-handed. Attacking ants that have room may pick up dropped beans. A group of ants carrying the maximum number of beans cannot attack.

WINNING: The game ends when all the beans on the board have been collected and stored in the nests. The player who collects the most beans wins.

TEAM PLAYER

Working in a group is often useful, but it can also be challenging.

2 In the game Forage on the previous page, what are some advantages of being in a large group?

_____

_____

_____

3 In the game Forage, what are some disadvantages of being in a large group?

_____

_____

_____

4 In the game Forage, what group size did you use most often when moving around the board? Explain.

_____

_____

_____

✏️ **JOURNAL:**

5 Think about a time when you had to work together in a group of other students or family members. Describe what you like or dislike about working in a group.

Lions are the only big cats that live in large groups. Other big cats like jaguars, leopards, and pumas live and hunt alone. What advantages and disadvantages do lions have compared with other big cats?

For each statement, mark whether it is an advantage or a disadvantage of living in a group.

6 Lions work in groups to hunt large prey like wildebeests and zebras.
◯ Advantage   ◯ Disadvantage

7 After a hunt, a lion must share most of its food with others in the group.
◯ Advantage   ◯ Disadvantage

8 Within a group of lions, some adults care for the cubs while others hunt.
◯ Advantage   ◯ Disadvantage

A lion's prey often live in large groups too. For prey animals like zebras, what are some benefits and challenges of living in a large group?

9 A large group of zebras is ◯ more ◯ less likely to be found by a hungry lion.

10 A lion is chasing a herd of zebras. Is one particular zebra safer in a small herd or a large herd? Explain.

_____

_____

_____

_____

WORKS WELL WITH OTHERS

Nope, this bird isn't sharing a secret with its friend. It's removing pests like flies and ticks from the impala's fur. Even animals from different species can benefit one another by working together.

11 Decorator crabs wear pieces of their habitat, including rocks, seaweed, or corals, on their shell. Corals are animals that capture small bits of food using tiny tentacles. Corals cannot move around on their own as adults and will stay attached to the crab.

In this relationship, the crab benefits from

- ○ **More food**
- ○ **Camouflage**
- ○ **Greater speed**

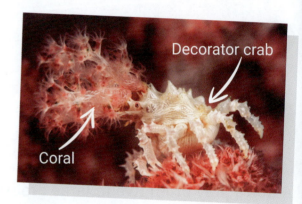

Decorator crab

Coral

12 Can you name a benefit that a coral may have by riding on the back of a decorator crab?

_____

_____

13 Can you think of a challenge either of these animals might face by living together?

_____

_____

CO-OP MODE

Is it true that two animals that eat the same foods always compete with each other? Sometimes animals in nature don't behave the way we expect.

14 Mongooses hunt for insects on the ground and some insects will fly away. Hornbills follow mongooses and snatch insects from the air that mongooses can't reach.

Which animal benefits from the mongooses stirring up insects into the air?

Beetle

Mongoose

Hornbill

15 Hornbills in the air can see for long distances and will call out when they see predators, like jackals. Mongooses recognize a hornbill's call and will hide underground.

Which animal benefits from the hornbill's call?

Mongoose

Hornbill

Jackal

16 Which sentence best describes the relationship between hornbills and mongooses?

- ○ Hornbills are predators and mongooses are prey.
- ○ Hornbills and mongooses help each other survive.
- ○ Hornbills and mongooses fight over food.

MVPs (MOST VALUABLE PLANTS)

Plants form relationships with many other kinds of living things, even us!

Plants can break down water and release oxygen into the air. Animals breathe in the oxygen made by plants.

Seeds often pass through an animal unharmed. There may be live seeds in animals' droppings.

Static charge helps pollen stick to the legs of bees and butterflies, just like a wool sock in the dryer might stick to a sweater.

Use the image on the previous page to help you choose the correct letter or letters that belong in the boxes below. You may use each letter more than once.

 A

 B

 C

 D

17 Which three animals are using a plant for food?

18 Which animal is helping a plant by spreading its pollen from one flower to another?

19 Which two animals are helping to spread the seeds of a plant to a new location?

20 Which of these animals use plants for shelter?

21 Flowers contain a sweet liquid called nectar. When bees drink nectar, they also move pollen from flower to flower. (See page 244 for more.) Explain how both the plant and the animal benefit in this relationship.

_____

_____

_____

DISCUSSION:

22 Pine tree seeds form inside of cones, not flowers. These trees do not rely on bees to move their pollen. Instead, pine tree pollen travels from cone to cone in the wind. Do you think a bee or the wind is more likely to get a plant's pollen to the right place? Explain.

DR. HAU

TEAMWORK II

Did you have fun working together, little monsters?

Yep!

Alex is great at making graphs, so he made all the graphs.

Lizzie can fly, so she was able to take some great close-ups of elefinches and their nests.

Grogg makes great drawings.

He drew all of these examples of how elefinches can help their group.

And Winnie made sure everything looked great when we put it all together.

RESEARCH PROJECT

Symbiosis (sim-bye-oh-sis) is a very close relationship between two different kinds of living things where one or both benefit. Let's research to learn more.

PRACTICE:

In the symbiotic relationships that follow, both living things benefit. Use the internet to discover how each partner helps the other and describe their relationship in the space given.

Ex.

CLOWNFISH

ANEMONE

Only a few animals don't mind an anemone's stinging tentacles. One such animal is the clownfish. Clownfish live very near anemones to protect themselves and their eggs from predators. Anemones benefit too because the clownfish helps clean the anemone and scare away fish that eat anemones.

RESEARCH:

23 *Search for "Pacific cleaner shrimp and moray eel symbiosis"*

PACIFIC CLEANER SHRIMP

MORAY EEL

24 *Search for "Dotted humming frog and tarantula symbiosis"*

_____

_____

_____

_____

_____

_____

DOTTED HUMMING FROG

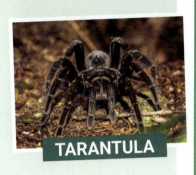

TARANTULA

25 *Search for "Acacia ant and bull's horn acacia tree symbiosis"*

_____

_____

_____

_____

_____

_____

ACACIA ANT

BULL'S HORN ACACIA TREE

HINTS
AND STRATEGIES

Use these hints and strategies for a nudge in the right direction on any of the starred problems or games in the book.

Chapter 1: Predictions 12-31

5. Here are some tips that may help:

- Make sure the floor is smooth and flat.
- Make sure the ball is let go in the same place on the ramp each time.
- Make sure the ramp doesn't move at all between rolls.
- Students who don't have a marble should try using a few different types of balls to see whether some are easier to predict than others.

10. Students should build parts of their contraption separately and test each part on its own so they don't need to reset the entire contraption every time they test it.

There are hundreds of videos of cool chain reaction contraptions online, Search "Rube Goldberg machine" for ideas.

Chapter 2: Evidence 32-49

6. Students who have learned about negative numbers and fractions can ignore them for this question and make Grogg's claim just about whole numbers (0, 1, 2, 3, 4, ...).

When is Grogg's statement true? Consider the pairs of numbers listed below. Circle the larger value in each pair. When are they equal?

1+0 1×0	2+0 2×0	3+0 3×0	4+0 4×0
1+1 1×1	2+1 2×1	3+1 3×1	4+1 4×1
1+2 1×2	2+2 2×2	3+2 3×2	4+2 4×2
1+3 1×3	2+3 2×3	3+3 3×3	4+3 4×3
1+4 1×4	2+4 2×4	3+4 3×4	4+4 4×4

Notice any patterns?

22. Students should think about other objects that they've seen fall. For example, a balloon filled with water and a balloon filled with air. Does one of these balloons fall faster than the other, even if everything else about them (size, shape, color, etc.) is the same?

23. How does a parachute slow a falling object down on Earth? What is different about the Moon that might affect how a parachute works?

29. Think about Alex's and Winnie's claims about what affects how fast an object falls. Then, think about what has changed.

It may also help to think about the earlier question about cyclists: Why does riding in a line help cyclists ride faster?

Chapter 3: Cause and Effect 50-69

8. Sometimes when we are asked to answer a question, it helps to think of extreme examples. Instead of just 3 washers, imagine a pendulum with something much heavier, like a bowling ball. Would a pendulum with a bowling ball on the end take longer to stop than one with a paperclip?

12. One way to test this is to make two pendulums side-by-side that are the same. Pull them both back the same amount and release them at the same time. Do they swing together? If not, make adjustments until they swing together at the same speed when pulled back the same amount.

Then, pull one back a long way and one just a little and let them go at the same time. Watch for at least 15 or 20 swings. Do they swing at the same speed, or does one get ahead of the other? If possible, use slow-motion video to see which one is faster.

18. About how long did the pendulum on pages 64-65 usually take to make 10 full swings? About how long did Grogg's usually take? What does this tell us about how the pendulums compare?

Chapter 4: Engineering 70-85

19. Experiment with using different weights. Which kinds of weights help a pendulum swing the same way for the longest amount of time?

20. The pendulum will be between 12 and 14 feet long (365 to 427 centimeters). Make sure the pendulum length is easy to adjust between trials, and be safe!

21. Try using a larger weight that weighs about as much as 10 of the smaller weights. Keep the two strings the same length. Experiment with adjusting the weights and lengths and see what happens.

22. Make sure that both of the weights are the same and are hung from the same length of string. The string on top should be able swing a little when one or more of the pendulums swing.

Next, try hanging three pendulums from the same rope. Swing different combinations and watch what happens.

Chapter 5: Temperature & Precipitation   86-105

13. Look at the charts on the previous pages to find the temperatures of Seattle, Chicago, and Phoenix in April. What color is the map where these cities are?

14. Look at the charts on the previous pages to find the temperatures of Seattle, Chicago, and Phoenix in May. How do these May temperatures compare to the temperatures in April? How would this change the way the map looks in May?

28. How could one city have more total rainfall but fewer rainy days than another?

37. Which container catches the most rain? How does the size of the opening in each container compare to the size of the container below it?

38. Does one container catch more rain than the others? Can one container hold more rain than the others?

39. How does the size of the opening in each container compare to the size of the container below it?

Chapter 8: Sun Angle   142-157

28. If we stand in the shadow of a cloud or a skyscraper, can we see the Sun? (Don't look at the Sun—we don't have to look at it to know whether we can see it or not.)

31. Where on Earth does the Sun disappear for months at a time?

Chapter 9: The Atmosphere   157-175

11. Will pumping air into the bottle 'trick' the altimeter into thinking it is getting higher or lower in the atmosphere?

12. At what altitudes does the atmosphere squeeze the bag the most? The least?

19. How does removing the cap change the way the atmosphere pushes on the water in the bottle?

20. How is this problem related to the two problems before it?

21. What happens to the pressure of the gas inside a can when we spray some gas out of the can?

22. When we flip the bottle, a little water usually drops out, but no air escapes. What is true about the air pressure inside the bottle compared to the air pressure outside the bottle?

23. What is different about space that may change the way a suction cup works (or doesn't work)?

Chapter 10: Wind   176-195

9. Will the air in the bottle (but outside the balloon) shrink or grow? How will this affect the balloon?

10. Will the air in the bottle shrink or grow?

11. Will the air in the bottle shrink or grow?

21. For air to flow into one area, it usually has to push out some of the air that is already there. When air flows out of one area, air usually comes from somewhere else to replace it.

22. What happens to air that is cooled on the slope by the snow?

25. For air to flow into one area, it usually has to push out some of the air that is already there. When air flows out of one area, air usually comes from somewhere else to replace it.

26. Where will the air rise at night? Where will it fall?

28. What heats and cools faster: land, or water?

29. What causes the wind that lifts these paragliders to blow in from the ocean?

31. When air flows out of one area, air usually comes from somewhere else to replace it.

32. What causes air to rise in the atmosphere?

5. Is water the only thing that controls how much food will grow in a habitat?

17. How will an increase in the number of wolves most likely change the number of elk in Yellowstone park? What will happen next?

19-24. Strategies for Pyramid Puzzles

Connect each animal to what it eats.

Start with animals that have only one choice of food or predator, these may be a little easier to place than the others.

24. What combination of animals can eat all of the grass?

28. Is there a habitat type on page 204 where plants can't grow?

30. Think about what happened to beavers in Yellowstone park when elk began eating young willow trees. Instead of willow trees, pandas rely on bamboo as their major food source.

Chapter 12: Competition 216-233

6-13. Territory Puzzle Strategies

Cross out squares where there aren't enough resources for a critter to survive. Using this strategy, we can see there are only 3 possible squares where a critter can be placed in the puzzle below.

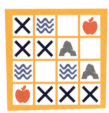

12. There are exactly four of each resource, so every resource must be used. Where could we place the critter that uses the water in the bottom-right corner?

13. There are exactly four of each resource, so every resource must be used. Where could we place the critter that uses the shelter in the bottom-left corner?

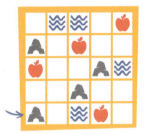

15. What is there more of up high that a plant might need to survive? What is there less of up high that a plant might want to avoid?

Pages 228-229: Canopy game strategies.

The more sun points we earn each turn, the more we can grow later. Players should look for ways to quickly increase the number of sun points they earn early in the game so they can grow more later.

As plants grow larger and closer together, growing above an opponent becomes more important. Players may try to shade their opponent's leaves from the sun to limit their sun points.

Players can also grow into spaces that will block their opponent's opportunities for growth or trap them. Branches may cross, so players may also find ways to escape.

Canopy is also fun to play using the pieces from games such as Go or Pente. Instead of filling in circles, students place tokens in empty squares of a gridded game board to grow their trees.

30. Will a larger number of owligators cause the number of chinchillipedes to go up or down? How will the change in the number of chinchillipedes affect the competition for food?

31. Consider what might eat the new animal and what the new animal eats.

Chapter 13: Adaptation 234-253

19. Both parts act as signals. What is the function of a signal?

31. The main body of a Venus fly trap is red and green. Use the information on page 248 and think about why each of these colors might be useful.

45. Colors can serve lots of purposes. Refer to page 248 for some examples of how an animal might benefit from its color.

Chapter 15: Protection 272-291

Pages 276-280: Predator and Prey strategies.

The ability with the highest rating isn't always the best choice against every opponent. Before choosing an action, consider what strengths and weaknesses your opponent may have.

For example, if you are a fox and your opponent is good at hiding (like a grasshopper), you might be better off fighting or chasing instead of detecting.

Not all of the matchups will make sense. For example, a snake might be able to detect a moose and injure it with a venomous bite, but a moose is much too large to be eaten by a snake.

11. Which berry bushes would not do well in a park that has deer?

12. Look for examples on pages 284-285.

24. Where is heat coming from - the cup, the snow cone, or the person's hand?

24. Where is there more heat in the snow cone example: inside or outside of the cup?

25. Would a sweater warm up a cold can of soda?

Chapter 16: Cooperation 292-311

Pages 300-301: Forage game strategies.

Experiment with different group sizes. Ants can split and regroup at any time during a turn. Sometimes this may mean some food is left behind.

For example, a group of four ants can carry five beans. Two ants can leave the group carrying two beans. The two ants who remain can later move with two beans (leaving one behind), or wait to be joined by more ants.

One strategy includes using a 'scout' ant who carries beans one at a time from a larger group back and forth to the hive.

10. Think about playing tag. If there is one person chasing a group, is it easier to avoid being tagged while playing with a group of 3 friends or 20?

12. What are some challenges the coral might have being stuck in one place its whole life? What benefits are there to moving around to different locations on the ocean floor?

13. What might happen if the needs of the coral are very different from the needs of the crab? What might happen if both animals need the same foods to survive?

20. Think of all the ways these animals may use plants for shelter from the elements like wind, rain, and sun.

SOLUTIONS

Find **hints** for selected problems marked with a star beginning on page 312.

1. Answers will vary. Generally, these types of trials can be surprisingly unpredictable, with the ball never seeming to land in the exact same spot twice. It's perfectly normal to miss a lot.

2. Answers may vary. Possible answers include:
   - The marble usually went straight, but often went too short or too far.
   - The marble usually went the right distance, but often missed to the left or right.
   - Every time I moved the cup, the marble went somewhere else.
   - The marble bounced off the cup a lot.

3. Answers may vary. Possible answers include:
   - The floor might have bumps that make the marble bounce in a different direction.
   - The marble might not be perfectly round.
   - The marble might roll a little faster or slower down the ramp sometimes.
   - The ramp might get bumped or moved.

4. Answers may vary. Possible answers include:
   - Make sure the floor is smooth and clean.
   - Try another marble.
   - Always place the marble the same way in the same place on the ramp, and release the marble the same way every time.
   - Make sure the ramp and the table don't move between rolls.

5. Answers may vary. Some combinations of ramp, floor, cup, and marble make it possible to bounce the marble in the cup lots of times. But, with a small cup or a rough floor, it might be very hard to get the marble in the cup at all. We recommend students try a different floor or a larger cup if they are having trouble.

6. For most students, it is easier to land the marble in the cup without a bounce. There's not much between the ramp and the cup that will change the direction the marble is going.

When the marble bounces off of the floor, small differences can make a big difference in where the marble bounces. Usually, the more events that happen, the harder the final results are to predict.

But, students may have problems with the ball bouncing in and out of the cup, or knocking the cup over. This is where filling the cup with rice or sand can be very helpful.

7. The higher we place the marble on the ramp, the more time it has to speed up and the farther from the table it will land.

8. The steeper the ramp, the faster the marble will go. Generally, this means it will land farther away from the table.

   But, if the ramp is too steep, the marble may not go as far. For example, if the ramp points nearly straight down, the ball will go straight down and land close to the table.

9. Answers may vary. In our experience, the marble has a more predictable bounce than most other balls. Balls that are very light or bouncy may be more difficult to predict.

10. Designs may vary.

11. Marble A has a steeper ramp and will roll faster than marble B. So, **marble A** will fall off the ramp and land in the water first.

12. Marbles C and D will speed up the same way down their ramps. Since marble C is higher, it will have farther to fall from the end of the ramp.
    So, **marble C** will land farther from the ramp.

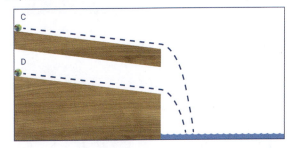

13. Marble F will speed up quickly on the steep part of the ramp, while marble E will begin slowly. By the time marble E speeds up on the steep part of its ramp, marble F will already be at or near the block. So, **marble F** will hit the block first.

14. Answers may vary. Two possible examples are shown below.

15. Answers may vary. Two possible examples are shown below.

16. Answers may vary. Two possible examples are shown below.

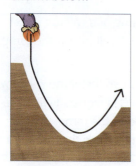 

17. Answers may vary. Two possible examples are shown below.

18. Answers may vary. Two possible examples are shown below.

19. Answers may vary. Two possible examples are shown below.

 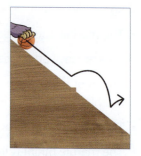

20. Answers may vary. Two possible examples are shown below.

21. The bean bag will be easiest to predict. Since it doesn't bounce, it will just land and stay where it hits the ground.

The football will be hardest to predict. Since it is not round, the direction it bounces will depend on what part of the ball hits the ground first. It's hard to know exactly what part of the football will hit the ground first.

If students have never seen a football bounce, find one and drop it a few times. Have them predict where it will end up when it stops bouncing.

22. The path of the ball landing on the smooth flat floor is easiest to predict. We expect it to bounce straight up or almost straight up until it rests.

The path of the ball landing on the jagged surface is hardest to predict. It could bounce in just about any direction, depending on exactly where it hits each time it lands.

Try dropping a bouncy ball on several different surfaces and trying to predict where it will end up when it stops bouncing.

Chapter 2: Evidence   32-49

1. Answers may vary. Any pair of whole numbers that are both greater than 2 will work. For example, 3+5=8 is less than 3×5=15.

2. Answers may vary. Common pairs will include a 1 or a 0. For example, 1+5=6 is greater than 1×5=5, and 0+5=5 is greater than 0×5=0. Students familiar with fractions and negative numbers may have many more examples.

3. In the previous problem, we found two numbers whose sum is larger than their product. Since multiplying two numbers doesn't *always* give a bigger result than adding them, Grogg's claim is not always true.

 Yes   ✔ No

4. Answers may vary. Any animal that doesn't have teeth works. For example, birds, insects, and animals without bones (like jellyfish and worms) are good examples of animals that do not have teeth.

5. Any rock that floats would be a counterexample. Some volcanic rocks float because they are filled with pockets of air!

6. Discussions may vary. Students may suggest replacing the word "always" with a word like "usually," which is true if we don't include fractions and negative numbers, and a good start for students who are only familiar with positive integers. But, we can do better.

Students may notice that Grogg's statement doesn't work when numbers like 0 and 1 are included. So, they may suggest that Grogg change

his claim to only include whole numbers that are larger than 1:

"Multiplying two whole numbers that are larger than 1 always gives a bigger result than adding them."

But, since 2+2=2×2, the claim above is still not true. A good revised claim is,

"Multiplying two numbers that are larger than 2 always gives a bigger result than adding them."

Students may have found other creative ways to revise Grogg's claim to make it true.

7. Alex knows that the toy without a parachute falls faster than the toy that has one. The two toys weight about the same amount, so this is evidence that narrow objects fall faster than wide ones.

Winnie knows that the bowling ball will fall faster than the balloon. The two objects are about the same size, so this is evidence that heavy objects fall faster than light ones.

8. The **ball** is wider.

9. The **ball** should land first.

10. The **flat sheet** is heavier (it began as the bigger piece of paper).

11. The **crumpled sheet** should land first.

12. Alex claims that narrow objects fall faster than wider ones. But, the large ball falls faster than the smaller balloon. This test gives a counterexample to Alex's claim.

Winnie claims that heavy objects fall faster than light ones. But, the lighter (crumpled) sheet of paper falls faster than the heavier one. This test gives a counterexample to Winnie's claim.

13. A person falling with an open parachute pushes a lot more air out of the way as they fall than a person falling before their chute is open.

14. Riding in a line, the person in front pushes the air for the riders behind. If the riders were side-by-side, they would push a lot more air.

15. The crumpled bag falls faster.

Bag

 ✔
Crumpled Bag

16. The crumpled bag has less air resistance.

17. The bag with the apple falls faster.

Bag   Bag + Apple

18. The bag with the apple weighs more.

19. The apple falls faster.

Apple   Bag + Apple

In this case, the difference may be hard to notice. It may help to use something smaller like a grape, or to drop both from higher to see the difference.

20. The apple has less air resistance.

21. Answers may vary. Designs that work best are usually the ones that have to push the most air as they fall. Some may also glide, spin, or cartwheel on their way down.

22. Answers may vary. Students have seen examples where objects of similar shape and size but different weight fall differently. So, they probably aren't convinced that weight doesn't matter at all. That's good! Weight does matter for objects that are falling through air.

We explain more on the page that follows.

23. Since there is no air to push out of the way, a parachute would not slow down a falling object on the Moon.

24. The cup should fall faster.

25. The feather and cup should fall together when the feather is in the cup.

26. The feather falls faster in step 3.

27. The size, shape, and weight of the feather don't change.

28. The air resistance on the feather is less when it is in the cup. The cup pushes the air out of the way.

29. The feather falls faster when it is in the cup because it does not need to push air out of the way as it falls.

30. Answers may vary. Students may be surprised that the book doesn't fall away from the feather, even after making the claim in the previous problem.

Since the feather does not need to push air out of the way, it falls just as fast as the book. The results support the claim in the previous problem.

Chapter 3: Cause and Effect   50-69

1. Answers may vary. Students will probably guess that the pendulum on the crane (A) will take the most time, but may make other guesses.

2. Answers may vary. Examples of student suggestions include the weight, length, and how far the pendulum swings (for example, students may guess that the lamp pull doesn't go very far back and forth when it swings).

3. Answers may vary. Student answers will usually be between 12 and 16 seconds for arm-length (35-64 cm) pendulums.

4. Answers may vary. The goal of this activity is stopwatch practice.

5. Answers may vary. Students may guess any of the three options.

6. Results may vary. Students should get roughly the same times for both pendulums. (All ten times in the chart should be close, with less than a second difference between the longest and shortest times if recorded correctly.)

7. The swing times should be about the same for the lighter weight as the heavier weight.

8. Answers may vary. Students may guess correctly that the 3-washer pendulum will swing for longer. It's harder to stop a moving object that has more mass because of a property of matter called inertia.

   It may help students who guess that the 1-washer pendulum will swing for longer to consider lighter weights, or even no weights. They can test their guesses.

9. Answers may vary. Students may guess any of the three options.

10. Ten swings of the short (30 cm) pendulum should take about 11.1 seconds. Ten swings of the long (60 cm) pendulum should take about 15.6 seconds. Times will vary depending on how well students measure the lengths and the swing times.

11. The long pendulum takes longer to swing back and forth than the short one. This is probably not surprising to students if they answered the first question in the chapter correctly.

   - ✔ The longer string took longer
   - ○ The shorter string took longer
   - ○ They were about the same

12. Believe it or not, how far we pull a pendulum back has almost no effect on how long a pendulum takes to make 10 swings.

   The weight on a pendulum that makes big swings goes a long way in each swing, but it moves fast through the middle of each swing. A pendulum that makes small swings doesn't go very far, but it moves very slowly. So, big swings and small swings of the same pendulum take about the same amount of time.

   It's hard to tell the difference with a stopwatch, but bigger swings take a tiny bit longer than smaller swings. Students may be able to notice the difference by comparing very big swings to very small swings.

13. Increasing the amount of weight at the end of a pendulum...

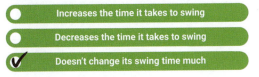

   - ○ Increases the time it takes to swing
   - ○ Decreases the time it takes to swing
   - ✔ Doesn't change its swing time much

14. Increasing the string length of a pendulum...

   - ✔ Increases the time it takes to swing
   - ○ Decreases the time it takes to swing
   - ○ Doesn't change its swing time much

15. If we change more than one thing during an investigation and get a new result, it's hard to know what cause created the effect. Changing one thing at a time lets us see what changes actually make a difference in the data we collect.

16. Answers may vary. Possible student answers include the materials used for the string and the weight, the size and shape of the weight, the way the string is attached, where the pendulum is located (on another planet?), or outside factors like wind, rain, or even air pressure.

17. All of Grogg's trials took very close to 19 seconds except for the circled trial below, which probably had an extra swing.

Time in Seconds (10 Full Swings)		
	YARN	FISHING LINE
TRIAL 1	18.96	19.05
TRIAL 2	18.89	19.14
TRIAL 3	19.03	18.97
TRIAL 4	(20.76)	19.02
TRIAL 5	19.08	19.12

18. Answers may vary. Most importantly, students should be able to tell whether the string on their pendulum was longer or shorter than the yarn and fishing line Grogg used.

   Students with shorter swing times should state that their pendulums were probably shorter in length than Grogg's. Students with longer swing times should state that their pendulums were probably longer than Grogg's. Grogg's pendulums were about 90 cm long.

19. Based on Grogg's data, string type has little or no effect on a pendulum's swing time. But, it would be good to test more string types to support this claim.

1. Changing a pendulum's length changes its swing time.

2. Changing the weight at the end of a pendulum does not really change its swing time unless changing the weight affects how long the pendulum is (when measured from the center of the weight to the top of the string). Besides length, most of the changes we can make have almost no effect on a pendulum's swing time.

3. Student answers may vary. Possible problems include the pendulum bumping into things, the string or weights coming detached, or difficulties changing the length and weights on the pendulum.

4. Student answers may vary. The goal here is to have students see various pendulums that might inspire design ideas. They may also come across some of the special pendulums we describe later in the chapter.

5. Answers will vary. Two examples are given on the following pages. Students should look for ways to attach their pendulums to a common object.

6. Answers will vary. Two examples are given on the following pages. Length adjustments should avoid tying and untying knots. For example, wrapping the string around something to shorten it as in Lizzie's design, or using a device like Grogg's keyring that can be used to adjust the string's length.

7. Answers will vary. Two examples are given on the following pages. Students should avoid tying and untying knots when changing weights. For example, weights can hook on to a looped string as in Lizzie's, or clip to a ring as in Grogg's.

8. Designs will vary. Students should label parts and include ideas about how their design will work. Students who are not interested in drawing may find other ways to describe their design like using photos and written descriptions.

9. Student answers will vary. Encourage students to think about the design challenges of portability and adjustability, and to look for potential problems with each design. Students may even

want to try building Grogg's or Lizzie's design to see what works well and what doesn't.

10. A pendulum that ticks 60 times in 60 seconds ticks once every second. So, 10 ticks will take 10 seconds.

11. An 80 cm pendulum takes about 9 seconds to make 10 ticks.

Students who get a different answer should:

- Check that their pendulum is 80 cm when measured from the top of the string to the center of the weight.

- Make sure that they are counting a swing in one direction as a tick (out and back is two ticks).

- Make sure that they don't count the first tick when they release the pendulum. It may help to say "zero" when the pendulum is released, then count each tick.

12. An 80 cm pendulum ticks 10 times in about 9 seconds. We want it to take 10 seconds. The longer a pendulum is, the slower it ticks. So, to make a pendulum that ticks 10 times in 10 seconds, it must be longer than 80 cm.

13. Answers will vary. A pendulum that ticks once every second is very close to 100 cm long. Below is an example of a table with the correct times for four different pendulum lengths. Students may make minor errors measuring lengths and times, but should arrive at a pendulum that is about 100 cm long after several trials.

Time for 10 Ticks	LENGTH (CM)	TIME (SEC)
TRIAL 1	90	9.57
TRIAL 2	95	9.83
TRIAL 3	100	10.08
TRIAL 4	98	9.98

In the past, scientists considered defining

the length of 1 meter based on the length of a pendulum that ticks once every second. But, swing times vary slightly based on several factors that make defining the meter this way difficult.

14. Student answers will vary. The table below gives approximate times based on pendulum lengths from 97 to 101 cm.

Time for 60 Ticks		
	LENGTH (CM)	TIME (SEC)
TRIAL 1	97	59.28
TRIAL 2	98	59.58
TRIAL 3	99	59.89
TRIAL 4	100	60.19
TRIAL 5	101	60.49

Students may make minor errors measuring lengths and times. As an extra challenge, students may try filming the pendulum swing in slow motion with a stopwatch in the video to see how close they can get to 1 second per swing.

15. Student answers may vary. Possible challenges include measurement difficulties, making small adjustments to the pendulum length, or getting the pendulum to swing the same way for 60 seconds (pendulums may slow down, spin, or begin going in circles instead of just back and forth). Students can try changing their design including the way the length is adjusted or the weight at the end.

16. Student answers may vary. Students should suggest testing indoors where there is no wind.

17. Student answers may vary. Grogg's pendulum weight probably isn't heavy enough to keep swinging for very long. Or, air resistance is slowing his pendulum down too much. A good suggestion is using a smaller but heavier weight at the end.

18. Student answers may vary. Lizzie should be able to make smaller adjustments to her pendulum's length. A difference of 2 cm will change the time it takes a pendulum to swing 60 times by a little more than half a second, so it may be hard to get exactly 60 seconds for 60 ticks. (But, she can get pretty close.)

19. Student designs may vary. Having a strong, thin string with a small but heavy weight at the end will help the pendulum swing the same way for a long time.

20. A pendulum that ticks once every two seconds will be a little less than 4 meters long. That's 4 times longer than a pendulum that ticks once every second.

We can square (multiply a number by itself) the time that it takes a pendulum to tick in seconds to get a good guess of the pendulum's length in meters. For example, a pendulum that ticks once every 3 seconds is about 3×3=9 meters long. A pendulum that ticks once every 10 seconds, it would have to be about 10×10=100 meters (almost 330 feet) long!

21. When we swing the small pendulum, it causes the much larger pendulum above it to swing. With the right weights and lengths, the pendulums create some interesting patterns at they swing, with the heavy weight stopping and starting as the small one swings below it. Play with different weight and string length variations to see what happens.

22. Swinging one of the pendulums causes the string it's attached to swing slightly. Eventually, the energy transfers to the pendulum that was not swinging. The pendulums will take turns swinging. Search the web for "coupled pendulums" to find video demonstrations of these cool pendulums.

1. Answers will depend on location.

    For the next 6 problems, answers should be within a couple degrees of the answer shown.

2. 59°F. Cool, Comfortable, Mild, or similar.

3. 30°C. Very Warm, Hot, or similar.

4. 23°F. Cold, Very Cold, Freezing, or similar.

5. 5°C. Cold, Chilly, or similar.

6. -20°C. Extremely Cold, Frigid, or similar.

7. 68°F. Comfortable, Nice, Pleasant, or similar.

8. Seattle: 73°F Chicago: 81°F Phoenix: 105°F

9. Seattle: 47°F Chicago: 32°F Phoenix: 68°F

10. Seattle: 5 Chicago: 4 Phoenix: 5

11. Answers will vary depending on location.

12. Discussions will vary. Students should familiarize themselves with temperatures that feel hot, cold, and comfortable to them.

13. The lightest blue stands for temperatures in the 30's and we count up by 10 for each color. Students unfamiliar with temperatures in the U.S. can look at the data on the previous pages to see April temperatures in Seattle, Chicago, and Phoenix.

    Students may be 10 degrees higher or lower on all of these temperatures.

14. Looking at the data on the previous page, we can see that Seattle, Chicago, and Phoenix all get about 10 degrees warmer from April to May. The map for May looks very similar to this one, but each region is one color warmer—the gray region becomes yellow, yellow becomes orange, etc.

15. Answers will vary. Most places are warmest in the early afternoon and coldest around sunrise. But, temperatures depend on lots of things, so this is not always true.

16. Answers will depend on location.

17. Answers will depend on location.

18. Answers will vary.

19. Container A holds more water than B, which holds more water than C. Even though the containers are filled to the same depth, container A is the widest and therefore holds the most water.

    Consider how many drops of rain fall into each container as it rains. The container with the largest opening collects the most rain.

20. Alex's pool should have collected about 2 inches of water. So, the water level will rise by 2 inches. (For most places, 2 inches is a lot of rain.)

21. Much more water fell in Alex's pool since it is so much bigger.

22. We look at the depth of the water in each container. The water is deeper in Lizzie's container, so Lizzie's house got more rain.

23. Answers may vary. Students may be surprised to see how little rain falls. It takes 3-9 hours of moderate rainfall just to get 1 inch (2.5 cm) of rain.

24. Answers will depend on location.

25. Answers will depend on location.

26. Answers will depend on location.

27. New Orleans gets more rainfall, but Seattle has many more rainy days.

28. Answers may vary, but the main idea should be that on days when it rains in New Orleans, it rains a lot more than on days when it rains in Seattle.

    Students may suggest that Seattle has lots of light rainfall, while New Orleans has heavy downpours that don't happen as often (which is true).

29. Student drawings will vary depending on the size of the cup, car, and pan. In most cases, the water in the pan will be very shallow and only part of the way up the tires on the car.

30. Student drawings will vary. In most cases, the car will be completely under water, with the water usually much deeper than the car is tall.

31. Answers may vary. With an absorbent towel, no water will pour from the pan; it will all stay in the

towel. With a less absorbent towel, some water may pour from the pan into the bowl, but much less than without the towel.

In areas with lots of plants and good soil, the plants and soil act like a towel, soaking up lots of rain and preventing floods. Areas with hard ground and pavement are more likely to flood because the water is not absorbed by the ground and instead flows quickly into low areas.

32. The tall grass (and the soil it grows in) will soak up and hold a lot of water, just like a towel.

    The brick pavers soak up almost no water.

    Loose soil and paving blocks are somewhere in-between.

Brick Pavers   Tall Grass   Loose Soil   Paving Blocks

33. Water will get trapped in the leaves, twigs, roots, and soil beneath the dense plants, slowing its flow.

    Water will quickly flow across the sidewalk.

Dense Plants   Grass Lawn   Concrete Sidewalk   Packed Dirt

34. Flooding happens when water collects on the surface, usually in low areas.

    Water that is absorbed or flows slowly from where it falls doesn't have a chance to collect and cause flooding.

35. Answers may vary. Any surface that is replaced with one that absorbs more water or slows its flow can reduce flooding. For example, replacing packed dirt with loose soil or grass, or replacing brick pavers with paving blocks like the ones shown will help the ground absorb more of the rain where it falls.

    Some replacements wouldn't make much sense. For example, replacing sidewalk with dense plants or tall grass would make it less useful for walking.

36. Answers may vary. Green roofs absorb and collect water so the rain that falls on the roof does not flow off and contribute to flooding.

    Other benefits include making the roof cooler and creating a nice space for people and birds.

    Possible problems include increased weight and installation cost. Surprisingly, green roofs tend to last longer and are no more leaky than many more traditional roofs.

37. Container A will fill up first and C will fill up last.

    This is because of the size of the openings. Consider the three containers below, whose openings are the same size as containers A, B, and C. As it rains, the depth of the rain will rise to the same level in all three containers. But, the largest container on the left holds much more water than the smallest one on the right.

Each container below will collect the same amount of rain as the container above that has the same-sized opening.

Imagine trying to pour the water from each container above into the container with the same-sized opening below to help see why container A will be first to fill up.

38. Container A will fill up first and B will fill up last. All three containers have the same-sized opening, so they will collect the same amount of rain. But, container A is the smallest so it will fill up first. B is the largest, so it will fill up last.

39. Container A will fill up first and C will fill up last.

All three containers hold about the same amount of water, but have different-sized openings.

Container A has an opening that is larger than the rest of the container. So, it will fill faster than a container with straight sides like container B.

Container C has an opening that is smaller than the rest of the container. So, it will fill up slower than container B.

40. As noted, 1 centimeter of rain falls as 10 or more centimeters of snow. So, snow takes about 10 times more space in the bucket than rain. We circle D, since it looks like the snow is about 10 times deeper in the bucket than the line at D.

41. Salt is too small and the soccer ball is too large. Hail stones can be the size of a marble, golf ball, or even a tennis ball!

Salt Grain   Glass Marble   Golf Ball   Tennis Ball   Soccer Ball

Chapter 6: Graphs   106-123

1. Fewest: Portland. Most: Las Vegas.

2. Las Vegas has 3 more suns which stand for 20+20+20=60 more days of sun in Las Vegas than San Diego.

3. All three statements are true in a normal year.

✔ Chicago has more sunny days than Portland.

✔ San Diego has about 20 more sunny days than Kansas City.

✔ Las Vegas has about twice as many sunny days as Tampa.

4. Half a Sun would stand for 10 days.

5. Chicago and Portland.

6. We fill the chart as shown.

Sunny Days Each Year						
City	Chicago	KansasCity	Las Vegas	Portland	San Diego	Tampa
Sunny Days	84	120	209	68	146	101
Missing Days	4	0	9	8	6	1

7. Answers may vary. Students may recognize that there are too many lightning bolts to count quickly. It is easy to see that Tampa has the most storms,

but not as easy to figure out about how many days have thunderstorms in each city.

8. The stacks of lightning bolts are the same height. It would be better if the lightning bolts were spaced the same way in each stack.

A better graph for problems 7 and 8 would look like the one below.

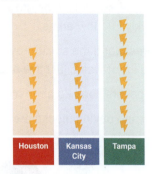

Houston   Kansas City   Tampa

9. Boston receives the most snow (about 50 inches per year). The graph on the right makes this clear. The graph on the left uses different-sized snowflakes and the stacks of pictures are all the same height. The bigger snowflakes may even suggest that New York gets the most snow.

10. All of the pictures should be the same size and spaced the same way.

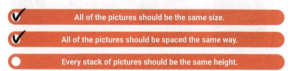

✔ All of the pictures should be the same size.

✔ All of the pictures should be spaced the same way.

○ Every stack of pictures should be the same height.

11. The snowfall graph on the right side of page 117 is the best of these four. The pictures are the same size and spaced the same way. It is easy to tell which city gets the most snow and easy to tell about how much snow falls in each city every year.

12. Answers may vary.

Some patterns students might mention are:

Both cities are warmest in July and August and coolest in December and January, with temperatures rising and falling between.

Chicago is hotter in summer and colder in winter than Seattle. In other words, temperatures don't change as much in Seattle as in Chicago.

These patterns are easier to notice in the graph, but the table makes it easier to see exact temperatures for both cities.

13. March: 52°F   June: 66°F
September: 67°F   December: 46°F

14. Chicago has both the hottest month (July, 82°F) and the coldest month (January, 32°F).

15. Seattle: August (73°F)   Chicago: July (82°F)

16. Chicago gets hotter in summer and colder in winter than Seattle. So, Chicago has larger temperature changes.

17. Answers may vary. Comparisons and patterns are often easier to see in a graph, but specific numerical information is often easier to see in a table.

So, finding the monthly high temperatures for Seattle is probably easier from the table. Finding the city with the hottest and coldest month is easier by looking at the graph. Finding the hottest month for each city is easy on the graph, but when months are almost the same temperature, it's good to check the table to be sure.

18. Mostly Cloudy: 120   Mostly Sunny: 115

19. The top statement is true.

✔ Denver has about the same number of cloudy and sunny days each year.

◯ Denver has about twice as many cloudy days as sunny days each year.

20. The graph on the left suggests that Denver has many more cloudy days than sunny days. This is because the number of days per year on the left side of the graph does not begin at 0. The graph on the right does a better job of showing that Denver has about the same number of cloudy days as sunny days.

Notice that the numbers don't add up to 365. Some days are not counted as mostly cloudy or mostly sunny.

21. The graph on the right suggests that it rains twice as much in Dallas as in San Jose. This is misleading because San Jose gets about 60 rainy days, and Dallas gets about 80, which is much less than twice as many rainy days.

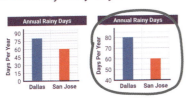

22. The graph on the right does a better job of showing the changing temperatures. On the left, the bars all look about the same height and it's hard to tell what temperature each bar stands for.

On the right, we can see that high temperatures in January are near 74°F and in July they are about 87°F.

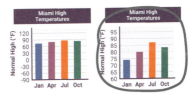

23. The graph shows that Vail gets about 350-355 inches of snow in a normal year, and Jackson Hole gets 365-370 inches. These amounts are similar, but the graph makes it look like Jackson Hole gets about twice as much snow as Vail, which is misleading.

24. Houston: 61   Memphis: 53   Orlando: 82

25. It storms more than 50 days in a normal year in Memphis. The friend may have been misled because the bar for Memphis is so much smaller than for Houston or Orlando.

26. The number of storm days on the graph doesn't go below 50. So, New York City doesn't have enough storm days to have a bar on the graph.

27. The graph should look similar to the one below.

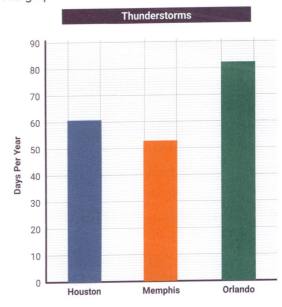

28. Graphs may vary.

1. Answers may vary. Day length in New York City changes gradually through the year. Days are longest in June at the start of summer and shortest in December at the start of winter.

2. June

3. December

4. March and September (October is close, too).

5. The seasons in Santiago are opposite the seasons in New York City. When it's summer in New York City, it's winter in Santiago.

   In New York City, the longest days are in June and the shortest days are in December. In Santiago, the longest days are in December and the shortest days are in June.

   But, the months that have 12-hour days are the same (March and September).

6. Buenos Aires and Punta Arenas have their longest days in December.

7. Havana never has days that are longer than 14 hours.

8. Punta Arenas has the longest summer days. Montreal is next.
   Buenos Aires has shorter summer days.
   Havana has the shortest summer days.

9. Punta Arenas has the largest difference in day length.

10. Havana has the smallest difference in day length.

11. The farther a city is from the equator, the bigger the changes in day length through the year.

12. In June, we see more of the top half of the orange. This is the northern hemihere on a globe.

13. In June, we see more of Seville, since Seville is on the top half of the orange.

14. In June, we never see the very bottom of the orange where the pencil sticks out. This is the South Pole on a globe.

15. In December, all of the answers are reversed. We see more of the bottom half of the orange, which means we will see more of Auckland. And we will

never see the very top of the orange, which is the North Pole.

16. In June, New York City spends more time in the light.

17. In June, it's dark all day long since the Sun does not shine on the South Pole in June.

18. In December, New York City spends more time in the dark.

19. In December near the South Pole, it is always light out since the South Pole is on the light side of the planet.

20. Summer has the highest temperatures in Rome.

21. Winter has the lowest temperatures in Rome.

22. Temperatures increase every month in spring.

23. Seasons in Sydney are in the opposite months as those in Rome. But, the answers to the previous questions are all the same. Summer has the highest temperatures, winter has the lowest temperatures, and temperatures rise in spring.

   Students may also notice that summer temperatures are hotter in Rome than in Sydney, and winter temperatures are colder in Rome than in Sydney.

24. The Sun stays up all month in May, June, July, and August in Longyearbyen.

25. The Sun rises in February after months of darkness.

26. It's dark all month in Longyearbyen during November, December, and January. During these months, the Sun is shining on the opposite side of the globe. So, it must be light during these months in areas near the South Pole.

27. One of the biggest differences between the seasons is the temperature change. Since temperatures in Singapore hardly change at all, people there probably don't experience the four seasons the way people do outside the tropical zone.

28. Three of the four places described can be found in the tropics.

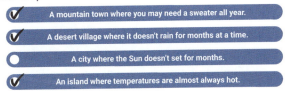

- ✓ A mountain town where you may need a sweater all year.
- ✓ A desert village where it doesn't rain for months at a time.
- ○ A city where the Sun doesn't set for months.
- ✓ An island where temperatures are almost always hot.

29. Answers may vary. Some possibilities:

- Students may prefer the heat or the cold.
- Students may want to avoid the months-long darkness near the poles.
- Students may find one of the tropical locations on the previous page appealing.
- Students may think it would be neat to experience endless summer days near the poles.
- Students may prefer the poles in summer and the tropics in winter.

Chapter 8: Sun Angle   142-157

1. The circle should appear brighter when the flashlight is straight above it, pointing down.

2. The light spreads over a wider area when the flashlight is at an angle.

3. The Sun is highest in the sky in the middle of the day (around lunchtime for many). This is the time when it provides the most heat and light.

(The Earth takes a while to heat up and to cool down, so the hottest part of the day is often several hours after the Sun is highest in the sky.)

4. The Sun is hottest and brightest at C where the Sun hits the surface straight on.

Section C is smallest but gets just as much Sun as the other areas (A, B, D, and E), which are larger. In those areas, the Sun hits the surface at an angle just like our flashlight when it was tilted.

5. We can count the 'rays' to see how many strike each section.
A is smallest with 3 rays.
D is next with 5 rays.
B is next with 6 rays.
C is largest with 7 rays.

6. The North Pole gets sunshine all day, but the Sun is at a very low angle. So, the Sun doesn't give as much light and energy as it does in areas where it is higher in the sky.

The North Pole was also dark and very cold for many months in winter when there was no sun at all. It takes a long time to heat up when the Sun finally rises.

7. Region D gets the same amount of sun in December as C does in June. The Sun is almost directly overhead.

8. We compare the numbers of rays in June to December.
Region A gets 3 more rays in June.
Region B gets 4 more rays in June.
Region C gets 2 more rays in June.
Region D gets 2 fewer rays in June.
So, the biggest difference is in region B.
(Think about the major differences between summer and winter in parts of the northern U.S. which are in region B.)

9. We count the rays.
Region D gets 7 rays in December, which is as much as regions C (5), B (2), and A (0) combined.

10. In June, region E (near the equator) will get slightly more sun than region D but less sun than region C.

In December, region E (near the equator) will get slightly more sun than region C but less sun than region D.

11. Person B will look straight up to see the Sun high in the sky. Person A will see the Sun low on the horizon. Person C is on the dark side of Earth and will see the night sky.

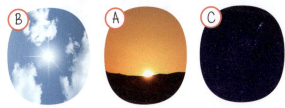

12. We shade the region where the Sun gets directly overhead.

13. We shade the region where the Sun gets directly overhead.

14. We shade the region that the Sun doesn't reach, near the North Pole.

15. Shadow A.

16. Shadow C.

17. Shadow B.

18. We draw an arrow pointing in the opposite direction of the shadows.

19. The shadow is almost directly below the cloud. So, the photo was probably taken in the middle of the day, around noon.

20. We draw an arrow pointing away from the dark side of the Moon.

21. North of the equator (outside the tropics), the Sun is highest in June and lowest in December. We label the boxes from top to bottom:

June

September

December

22. South of the equator (outside the tropics), the Sun is highest in December and lowest in June. So, those two box labels would be reversed.

23. The bright streak in the photo on the right is the Moon. In the bottom picture, the light streaks are headlights of automobiles along a road.

24. Answers may vary.
There should be a curved, faded path across the paper made where the sunlight faded the paper. It should be most faded in the middle, when the Sun was highest and the circle of light shining on the paper was brightest.

25. We check the two items below.

26. Earth's tilt causes the change in seasons which makes the Sun higher and days longer in summer.

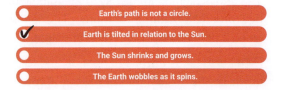

27. December.

28. The shadows of clouds and skyscrapers are the parts of the Earth that are blocked from the Sun. So, if the Sun had eyes, it wouldn't be able to see the shadows behind objects it shines on.

29. Shadows are shortest when the Sun is highest in the sky. In places outside of the tropics, this is in the middle of the day (usually around noon, depending on location) at the start of summer (around June 21st north of the equator, or December 21st south of the equator).

On the equator, the Sun is directly overhead twice each year in March and in September. Other tropical locations also have midday sun that is directly overhead twice each year.

30. The solar panels are tilted so that the sun shines directly at them when it is high in the sky. This way, they can collect as much energy as possible.

North of the equator, panels are tilted towards the south, and south of the equator, panels are tilted north. The farther we are from the equator, the more tilted the panels should be. Some panels even adjust automatically to point at the Sun.

31. This is how the Sun appears in the sky when viewed from the North Pole!

Chapter 9: The Atmosphere   157-175

1. People near the bottom of the tower carry the heaviest load and have the most pressure on their shoulders. They carry not just the weight of the people standing on their shoulders, but also the weight of the people higher up in the tower.

2. The bottom box is likely to break first because it carries the weight of all the boxes above it.

3. The marshmallows near the bottom are being pressed by the weight of the marshmallows above. So, they will be more tightly squeezed and squished together compared to the marshmallows on top.

4. When we squeeze a marshmallow, it will push back and fill any available space around it. A ping pong ball at the bottom of our marshmallow-filled tube will be squeezed from all sides (even below) by the marshmallows all around it that are being squeezed by each other.

5. Answers may vary.

6. The marshmallows get smaller.

If this is not the case, the bottle is probably not sealed well enough or the pump is not working.

7. The marshmallows grow back to about the same size they were before the air was pumped in.

8. The marshmallows stay about the same shape. The air presses the marshmallows from all sides equally. Depending on the marshmallows used, the outer layer may crease, crack, or wrinkle, and the marshmallows may get a little lumpy, but the overall shape stays about the same.

9. The marshmallows will grow if we remove air from the bottle (the same way that they grow when some air is let out of the pressurized bottle).

10. The air pressure decreases as we get higher. So, as air pressure drops, an altimeter will show the altitude as increasing.

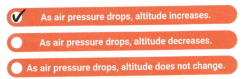

✓ As air pressure drops, altitude increases.

○ As air pressure drops, altitude decreases.

○ As air pressure drops, altitude does not change.

11. We can 'trick' an altimeter by changing the pressure of the air around it. Since pressure gets higher when we are lower in the atmosphere, the increased pressure in the bottle causes the altitude reading to drop!

12. As we move up through the atmosphere, the air pressure drops. Since the atmosphere doesn't squeeze the chip bag as much at the top of the mountain as at the bottom, the chip bag appears to inflate as we climb, even though no air is getting into the bag.

With lower pressure all around it, the air in the bag can take up more space. This makes it look like the chip bag is inflating, just like the marshmallows in the bottle when the pressure decreased.

13. The air that is coming out of the tire should feel cool.

14. The bottom of the tube where pressure is highest should feel warmer than the top of the tube.

15. When students pump air into the bottle, it should feel warmer. When the air is released, the bottle should cool.

Students may only feel this difference if air is added quickly and the bottle doesn't leak.

16. As we increase the pressure in the bottle, the temperature increases. When we decrease the pressure in the bottle, the temperature decreases.

17. In the atmosphere, pressure is highest down low and lowest up high.

Temperature is also (usually) highest down low and lowest up high.

As we move in the atmosphere, temperature and pressure both usually decrease as we get higher, and increase as we get lower. When air moves down and is "squeezed" under pressure, it heats up. When air moves up, it expands and cools down.

This is true for the bottom part of the atmosphere (where most of the air is). Much higher, where there is very little air, there are parts of the atmosphere where temperatures increase as we get higher.

18. Air pressure holds the water in the bottle.

    That's the short answer. But, it's actually the *difference* between the air pressure inside the bottle and outside the bottle that keeps the water in. When students first poke the hole, a small amount of water leaks out. This makes a little more room for the air inside the bottle. Nothing gets into the bottle to fill the space.

    Since the air in the bottle has more space to fill with the same amount of air, the pressure of the air inside the bottle is lower than the pressure outside the bottle. The extra air pressure outside the bottle is what holds the water in.

19. Opening the bottle lets air push on the water from above. So, the water pours from the bottle just like we'd expect it to.

20. The hole is there to let air in. Otherwise, the difference in air pressure inside and outside of the cup would make it harder for the cocoa to flow out of the cup (as we've seen in the two previous problems).

21. When we let air (or any other gas) out of a pressurized can, the pressure of the air in the can decreases. When the pressure of the air in the can decreases, so does its temperature. (This is because the air uses energy to push air out of the can. When a gas loses energy, it cools down.)

    So, the can feels cold when air is released and the pressure in the can decreases.

22. Similar to problem 17, air pressure keeps the ball 'stuck' to the bottle.

    When we flip the bottle, a little water drips out, leaving more space in the bottle for the air inside.

Since the air in the bottle has more space to fill with the same amount of air, the pressure of the air inside the bottle is lower than the pressure of the air outside the bottle. The extra air pressure outside the bottle is what holds the ball up.

23. The pressure from the weight of all the air in the atmosphere is what holds a suction cup to a smooth surface.

    But, if the suction cup is not sealed tightly against the surface, air can get in. That lets the air from the atmosphere press on all sides of the suction cup, not just the side that pushes it against the surface. So, the suction cup falls off.

    On a satellite in space, there's no air to press a suction cup against a smooth surface. So, suction cups won't work at all in space.

Chapter 10: Wind · 176-195

1. No. Once the cap is on, no air can get in or out.

2. The bottle shrinks a little and crumples in the cold water, then goes back to its original size in the hot water.

3. The air shrinks when it's cold and expands when it's warm. This is what causes the bottle to shrink and grow.

4. Hot air takes up more space. We can see this in the bottle experiment. When the bottle is in cold water, the air inside shrinks and the bottle crumples. When we return it to the hot water, the air inside expands and the bottle goes back to its original size.

5. No. With the cap off, the air could escape the bottle when we heat it, and enter the bottle when we cool it. So, the bottle would not shrink and grow with the cap off, the air would just move in and out.

6. Answers may vary. Students may predict that the balloon will inflate a little or a lot. Since hot air takes up more space, it makes sense to predict that the balloon will fill up when the bottle is heated.

7. Answers will vary. Students may predict that the balloon will shrink a lot or a little. Since cold air

takes up less space, it makes sense to predict that the balloon will shrink when the bottle is cooled.

8. Answers will vary. The balloon should grow when the bottle is heated and the air inside expands. The balloon should shrink when the bottle is cooled and the air inside shrinks.

9. Answers will vary. The air in the bottle shrinks when it is cooled. Since the air in the bottle (between the balloon and the inside of the bottle) takes up less space, the balloon will fill up a little. The pressure of the atmosphere fills the balloon to fill the extra space in the bottle.

10. Answers will vary. As the air in the bottle heats up, it will expand and begin to escape the bottle. Students should see bubbles come out of the hole in the cap.

11. Answers will vary. As the air in the bottle cools down, it will shrink. This will leave space inside the bottle for water to flow into. Some water will enter the bottle through the hole in the cap.

12. Answers will vary. Try using temperature differences to crush a soda can or pull an egg into a bottle. Or, find videos of others doing it online.

13. It's easiest to feel the heat from a candle directly above it, **30 cm above the flame**.

14. The hot air rises **up** above a heated pot.

15. Smoky air flows **up and out the top** of a chimney.

16. Smoke is heated and rises to the highest part of a room.

17. Near the surface, air is heated by land (and water) that have been warmed by the Sun. Air will be hottest right on the pool deck (which will also be hot). So, it would probably be easiest to keep cool in the lifeguard chair.

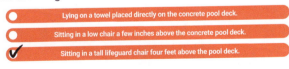

18. Hot air will flow **up until it is the same temperature as the surrounding air.**

19. The air near the heated surface of the mountain will be **warm** compared to the air around it.

20. The warm air in the circle will move up and out of the circle as shown in the diagram below.

21. The air below the circle will move in to fill the area left by the air moving up and out of the circle.

22. If the surface of a mountain is colder than the nearby air, it will cool the nearby air. This cooler, heavier air will slide down the mountain.

23. The air will rise above the sandy beach. (See below.)

24. The air will fall above the cooler ocean water. (See below.)

25. The breeze will blow from the ocean towards the sand. The air being pushed down from above moves to replace the rising air. (See below.)

26. At night, the land is cooler than the water. The air will rise above the water and fall above the sand. The causes the breeze to reverse direction, blowing from the sand to the sea. (See below.)

27. The equator is heated year-round by the Sun. This heated air tends to rise more than at other parts of the globe.

28. Since land heats and cools faster than water, the northern part of the globe (where there is more land) heats and cools faster than the southern part of the globe.

29. When the land is warmer than the water, air usually moves from the ocean towards the land as described on page 193. This is called a sea breeze. The sea breeze pushes against the cliffs, creating a wind that blows upward. Paraglider pilots can ride the rising air to soar high above the cliffs.

    This sea breeze is not as common in the morning, since the Sun hasn't had much time to heat the land. Sea breezes are usually strongest in the afternoon after the Sun has heated the land for most of the day.

    So, it is usually easiest to fly high above the cliffs in the **afternoon**.

30. Unlike air, water absorbs lots of sunlight. Since sunlight is absorbed in **the top part of the ocean**, this is where the water is usually warmest.

    Like air, cool water is heavier than warm water. So, cool water sinks and warm water rises. This is another reason it's generally colder at the bottom of the ocean than at the top.

    But, below about 4°C (about 39°F), water actually gets lighter as it cools. Ice floats, which is great because otherwise the oceans would all freeze from the bottom up!

31. If all the wind near the ground is blowing toward a field, it means that the air on the field is probably rising. When air rises on a field, the surrounding air rushes in to replace it.

32. Summer afternoons are when the ground is heated most by the Sun. The ground heats the nearby air, which rises high in the atmosphere. Moisture in the air can be seen as clouds when the air cools.

    In times when there is less energy from the Sun, clouds like these are less common.

Chapter 11: Habitats

1. The tundra is the driest of these habitats.

   ○ Grassland   ☑ Tundra   ○ Rainforest

2. The kelp forest is the warmest of these habitats.

   ☑ Kelp forest   ○ Ocean floor   ○ Tundra

3. No. Desert habitats have abundant sunlight but little food because the high temperatures cause water there to evaporate quickly. Without much water, most living things struggle to survive.

4. Answers may vary. Avoiding being eaten is a challenge living things face even in habitats with plenty of food and water.

5. Discussions may vary. The ocean floor has lots of water but less food than many other habitats.

   Students may also notice that cold habitats (tundra) have little food. Cold temperatures and lack of sunlight limits the ability of plants and algae to grow.

6. We fill in the missing animal names as shown.

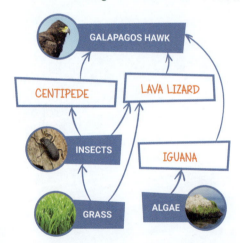

7. We fill in the missing animal names as shown.

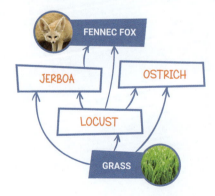

8. We fill in the missing animal names as shown.

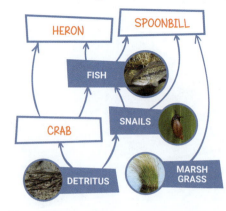

9. We fill in the missing animal names as shown.

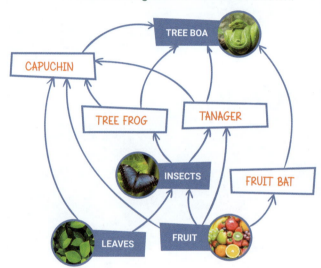

10. We fill in the missing animal names as shown.

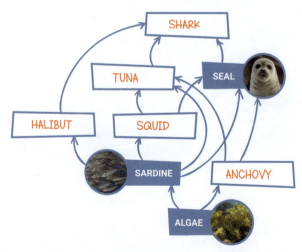

11. The number of elk **increased**.

12. The number of beavers **decreased**.

13. The number of willows **decreased**.

14. There were no longer any predators to eat the elk, so the number of elk increased.

15. The number of elk increased. More elk in the park were able to eat more willow trees.

16. Since wolves eat beavers, adding wolves might cause the number of beavers to decrease.

17. If wolves eat some of the elk, there will be fewer elk eating willows. The number of beavers may increase because there is more food for them to eat.

18. Findings may vary since wolf restoration is ongoing. Students will likely find that adding wolves back to Yellowstone park in 1996 has led to a decrease in elk and an increase in beavers.

In problems 19-24, letters in each row can be in any order.

19. We fill in the spaces as shown.

20. We fill in the spaces as shown.

21. We fill in the spaces as shown.

22. We fill in the spaces as shown.

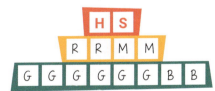

23. We fill in the spaces as shown.

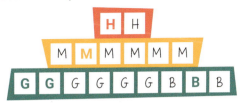

24. We fill in the spaces as shown.

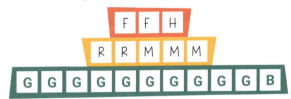

25. Twelve grass plots can feed 4 rabbits. Since one fox can eat a single rabbit, the greatest number of carnivores in this puzzle is **4** foxes. Twelve grass plots could also feed 6 mice, or 2 rabbits and 3 mice, but this is only enough to feed 3 carnivores.

26. There is no solution because 3 herbivores is not enough to feed 5 carnivores.

27. Answers may vary.

28. Answers may vary. Habitats with very little light have few plants and therefore few plant-eating animals. Much of the food on the deep ocean floor comes from animals that have died and drifted down. The deep sea habitat on page 204 is likely to have more carnivores than herbivores.

29. Answers may vary. Food supply and temperature extremes are two main challenges animals may face. For example, it is harder for some animals to find food and stay warm in winter.

30. An animal that only eats one type of food is at risk of starving if that food becomes scarce. Animals with many food choices are more likely to survive if their habitat changes.

31. Owls eat mice to survive. Mice eat plants that grow using energy from sunlight. Students may also point out other reasons that the Sun is important to life on Earth, such as warmth.

32. Stories will vary. Large forests (plants) might be visible from space using a powerful telescope. Intelligent aliens that send signals across space would most likely resemble animals. Microbes can live in almost any habitat and for this reason they may be more likely to survive on a harsh planet such as Mars.

1. Animals get energy, carbon, and minerals from their food.

   Animals living in the desert and other dry areas may also get water from their food.

2. Plants get their energy from sunlight, so they can't survive deep in the ocean.

3. Since hydroponic plants are not grown in the soil, their water must contain added minerals.

4. Mold is a fungus, and fungi get energy from foods.

   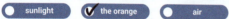

   The energy from an orange was made by a plant using sunlight, so the mold is *indirectly* getting its energy from sunlight.

5. Discussions may vary. Fungi and animals get most of their resources from the foods they eat. Plants get their resources directly from air and sunlight.

6. We solve the puzzle as shown.

7. We solve the puzzle as shown.

8. We solve the puzzle as shown.

9. We solve the puzzle as shown.

10. We solve the puzzle as shown.

11. We solve the puzzle as shown.

12. We solve the puzzle as shown.

13. We solve the puzzle as shown.

14. Discussions may vary. Animals often fight over resources. Hummingbirds might fight over access to flowers (food), water, mates, or to protect a nesting site.

15. Discussions may vary. Epiphytes that grow high off the ground have a better chance of getting sunlight. Being high up may also help them avoid getting eaten or make it easier for insects and small birds to visit them and spread their pollen.

16. Recall that unlike plants, fungi do not directly need sunlight and therefore don't compete.

   ○ Yes   ☑ No

17. The supply of plants available to eat in this area is much greater in the summer compared with the winter. Students should notice the level of water in the river has not changed much, so the supply of water here is similar in winter and summer.

18. We circled the first photo, which shows a greater demand for shrimp and algae. There is higher demand for food when there are a larger number of animals in the same place.

19. There are many trees available and there are also many birds that need to build nests.

   Supply: **HIGH**   Demand: **HIGH**

20. There is little water to drink in this dry savanna, but many thirsty animals are gathered in one place.

   Supply: **LOW**   Demand: **HIGH**

21. Many fish surround a single hungry shark.

   Supply: **HIGH**   Demand: **LOW**

22. There are very few plants or fruits to eat, but there don't seem to be many hungry animals around.

   Supply: **LOW**   Demand: **LOW**

23. Discussions may vary. Students may consider how the plants and animals in each photo compete for resources. Often the greatest competition exists when there is high demand but low supply, so the savanna habitat has a great deal of competition.

   Students may also argue that birds may compete for space to nest because some trees may be better than others. There could be unseen animals in the habitats that affect the amount of competition.

24. If only the supply of water decreases, then competition for water increases.

   ☑ higher   ○ lower   ○ about the same

25. If only demand for nest sites decreases, then competition for nest sites decreases.

   ○ higher   ☑ lower   ○ about the same

26. If only the supply of worms decreases, then competition for worms among newts will increase.

   ☑ higher   ○ lower   ○ about the same

27. If there are more animals eating fruit on the island, then competition will increase.

   ○ The bats there begin eating insects.   ☑ The number of bats increases.
   ○ The number of lemurs increases.   ☑ The lemurs there begin eating fruit.

28. Discussions may vary. Students might argue that as water in a habitat decreases, fewer plants will grow, resulting in increased competition for plants.

   Students could also argue that less water will result in fewer total living things in the habitat. A habitat with fewer animals would have less demand for food, causing competition to decrease.

29. Adding owligators to this habitat will most likely cause the number of chinchillipedes to decrease. If there are fewer chinchillipedes, competition among porcupumas for them will increase.

30. Adding owligators to this habitat will most likely cause the numbers of chinchillipedes to decrease. If there are fewer chinchillipedes, there will be a greater supply of pine-apples and competition among chinchillipedes will decrease.

31. Answers may vary. Students should consider what the Beast-world animals eat. Here are some example responses.

   A new animal in this habitat might be one that porcupumas eat. Porcupumas will have more food and this will reduce competition among porcupumas.

   If the new animal eats porcupumas, there will be fewer porcupumas. Decreasing demand reduces the competition for food among the remaining porcupumas.

1. A penguin's wings are more similar in shape to a sea lion's flippers.

2. A penguin's feet are more similar in shape to a sea lion's feet.

3. The inside of penguins' bones have tiny holes, more similar to sea lions' bones. These tiny holes are filled with meaty marrow.

Magpies, and all other flying birds, have large holes inside their bones filled with air.

Birds breathe harder when they fly, just like we do when we exercise. Scientists think the air pockets in birds' bones help them to get enough oxygen while they are flying. The air pockets also make the bones of flying birds much lighter.

4. Penguin chicks huddle together in groups to stay warm.

5. Slippery fish are easier to hold when you have a spiky tongue.

6. A fatty layer of blubber under their skin keeps penguin warm on land or in icy water.

7. If walking is too slow, penguins can slide on their bellies instead.

8. Penguins rub their feathers with oil (or preen) to stay waterproof.

9. Being waterproof is important because penguins spend a lot of time hunting for fish.

10. Students will choose their own tasks to complete. Results may vary, but we expect most tasks to take longer without using thumbs.

11. Mole hands are adapted for digging (C).

12. Bat hands are adapted for flying (B).

13. River otter hands are adapted for swimming (D).

14. Chameleon hands are adapted for climbing (A).

15. Answers will vary. Students should design a limb that helps them do a specific task.

16. Plant roots and animal mouths are both used for **taking in water**.

    Roots also take in some nutrients (like minerals) from the ground, but most of a plant's nutrients are made by the plant.

17. Plant leaves and animal mouths are both used for **getting food**.

    Plant's leaves allow them to make their own food, while animals eat either plants or other kinds of living things for their food.

18. Plant toxins and animal scales are both useful for **not getting eaten**.

19. Plant flowers and animal dewlaps are both for **using signals to communicate** with others.

    Flowers and dewlaps are also important parts related to **making offspring**. Seeds are formed within flowers and male lizards use their dewlaps to attract the attention of female lizards.

20. Switch the circled boxes to fix both sentences.

21. Switch the circled boxes to fix both sentences.

| Rattlesnakes are | intelligent; | they can use tools and solve puzzles. |
| Crows are | venomous; | their fangs inject poison into their prey. |

22. Switch the circled boxes to fix both sentences.

| Hedgehogs have | thick fur to | stay warm in freezing weather. |
| Polar bears have | sharp spikes to | protect them from being eaten. |

23. Switch the circled boxes to fix both sentences.

| Snails have hard shells | to protect them from | their parents. |
| Baby birds make peeping sounds | to communicate with | predators. |

24. Switch the circled boxes to fix both sentences.

| Coconut seeds | spread to new places | with strong smells. |
| Flowering plants | attract insects like bees | by floating across the ocean. |

25. Switch the circled boxes to fix both sentences.

| Horses' hooves are useful for | grabbing branches | on wide open plains. |
| Monkeys use their flexible tails for | running quickly | in tropical forests. |

26. Switch the circled boxes to fix both sentences.

| Cacti store water in their stems | so they don't dry out | in the icy tundra. |
| Cushion plant leaves turn purple | to absorb extra heat | in the desert. |

27. Switch the circled boxes to fix both sentences.

| Sharks' teeth | shoot out of their mouths at high speeds | to catch bugs. |
| Chameleons' long tongues | fall out constantly | and are replaced by new ones. |

28. Answers will vary. Each mixed up sentence should make sense when read, even if it says something silly. When one pair of phrases is swapped, both sentences should be facts about animal adaptations.

29. We circled the female bird in each pair below.

30. Most animals that eat insects want to avoid being stung by wasps. Looking like a wasp could help these sneaky insects frighten away predators.

31. The color green is a clue that Venus fly traps use sunlight to make sugars for energy.

Venus fly traps live in habitats that lack nutrients, like nitrogen. Instead of getting these resources from the soil, these plants get the nitrogen they need from the bodies of flies and other insects.

32. A — dead leaf butterfly
33. G — ornate ghost pipefish
34. D — thorn bug
35. H — copperhead snake
36. F — mossy frog
37. C — arctic fox
38. E — common octopus
39. B — ghost crab

40. Discussions may vary. Typically, camouflage helps a predator avoid being seen by its prey. Orchid mantises are predators that catch insects. Many insects eat plants and flowers, so the disguise can lure insects towards the mantis.

41. Answers will vary.

42. Answers will vary. Students should match the limbs of their creature to a reasonable habitat. For example, animals with hooves likely run on open plains. Animals with claws might climb trees. Animals with tentacles or fins likely live in the water. Animals with webbed feet likely spend some time on land and some time in the water.

43. Answers will vary. Students may decide whether their creature eats plants, meat, or both. Students should consider what body parts may be useful for gathering food.

44. Answers will vary. Students should correctly identify which parts of their creature are evidence that it needs protection from predators or weather. For example; hard shells, sharp spikes, and toxins are useful protection from predators. Thick fur and sleek scales might indicate that the creature lives in a very cold or very hot habitat.

45. Answers will vary. Students may choose to describe the advantages of their creature's color as a signal, camouflage, or a warning of toxicity. Creativity is encouraged!

Chapter 14: Feeding   254-271

1. Your incisors are most similar to the (B) fingernail clippers.

2. Your molars are most similar to the (A) nutcracker.

3. Your cuspids are most similar to the (C) knife.

4. A bobcat's molars have large sharp points while a kangaroo's molars are flatter with small ridges.

5. The kangaroo does not have **cuspids**.

6. The tapir most likely eats plants with its flat back teeth and large gap like the rhino.

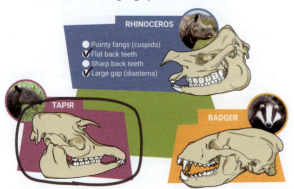

7. The river otter most likely eats meat with its pointy fangs and sharp back teeth like the seal.

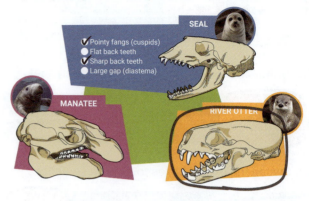

8. The tarsier most likely eats insects with its pointy fangs and sharp back teeth like the hedgehog.

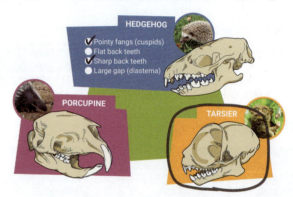

9. The wombat most likely eats plants with its flat back teeth and large gap like the rabbit.

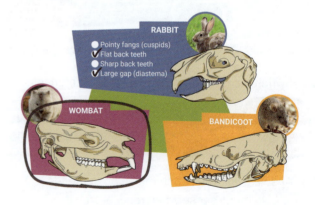

10. The musk deer most likely eats plants with its flat back teeth and large gap like the giraffe.

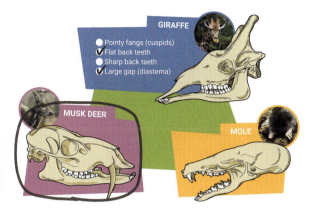

The three-foot-tall musk deer doesn't use its fangs for eating. They are used for protection, similar to the way most deer use horns or antlers.

11. The skulls of the kangaroo, rabbit, rhino, manatee, wombat, giraffe, and horse on pages 261-265 are evidence that most plant eating animals (herbivores) have similar teeth.

12. The skulls of the badger, bandicoot, opossum, bear, and monkey on pages 262-265 are evidence that most animals that eat both plants and meat (omnivores) have similar teeth.

13. The skulls of the bobcat, seal, otter, hedgehog, tarsier, and puma on pages 261-265 are evidence that most meat-eating animals (carnivores) have similar teeth.

Carnivores that only eat insects often have smaller cuspids that are similar in size to their other teeth.

14. This animal only has teeth in the back of its mouth.

15. Answers may vary. So far, we have observed that most animals with flat back teeth are plant-eaters (herbivores).

16. An aardvark uses its flat molars to crush insects.

17. Students may discover different facts about anteaters, pangolins, and echidnas. Some common features of their skulls are a long snout and no teeth. Without teeth, these animals grind food on the roof of their mouths with their tongues or swallow their meals whole. Here are some more interesting facts:

Echidnas are very unusual mammals that lay eggs. Echidnas have a single tooth that helps them hatch from their egg, but then they lose it.

A giant anteater uses its tongue to catch ants and termites. Its tongue is about two feet long and attaches to its body near the top of its ribs!

18. We match insect mouths to the following tools.

19. Pasta strainers (C) work as filters capturing pasta and letting the water drain out.

Chapter 15: Protection   272-291

1. Answers will vary. Students should correctly use the table to convert their speed in seconds to a speed in miles per hour. Most students should end up somewhere between the mouse (7 mph) and the squirrel (12 mph).

2. Answers will vary. Students should consider what adaptations animals might have to defend themselves, such as camouflage, flight, or armor (like the pangolin on page 273).

3. Answers may vary. Students will probably select A or C as most likely to get eaten and B as least likely to get eaten.

4. Answers may vary. Students should notice that some moths have markings that make them better camouflaged than others. Birds will eat the moths that they can see most easily, and moths with the best camouflage are most likely to survive.

5. Moth A, the lightest colored moth, is most likely to survive because it will camouflage best with the white birch trees.

6. Answers may vary. Students should consider how small size could help a fish survive. The fish may be better able to hide in small spaces or otherwise go unnoticed by the barracuda.

7. Answers may vary. Students should consider how large size could help a fish survive. The fish may be too large for the barracuda to eat.

8. **Bush A**, the plant without thorns is more likely to have its berries eaten by an animal, helping it spread its seeds.

9. **Bush C**, the plant with large thorns has the best protection against being eaten.

10. **Bush C**, the plant with large thorns, offers the best protection for mice from predators.

11. In the park with only deer, bush A will be completely eaten by the deer and bush C will not have its berries eaten to spread its seeds. So, a bush with small thorns (B) will do best in the park with only deer.

    In the park with mice and foxes, mice will be best protected by a bush with large thorns (C). They will eat the berries and spread the seeds.

    This leaves the bush with no thorns (A) for the park with only mice, who do not need protection and will eat and spread the seeds.

A park with only deer   A park with only mice   A park with mice and foxes

12. Yes, an adaptation can have more than one use. Students may find the following examples from pages 284-285, or come up with their own. A polar bear's white fur can be useful as camouflage and as protection from cold weather. Cactus needles give protection from being eaten and from drying out in the dry desert. Hiding in a burrow can help an animal stay cool and protect it from predators.

13. Answers will vary. Students should record the temperature every 10 minutes using either °C or °F.

14. Answers will vary. Students should subtract the starting temperature from the final temperature to find the total change after 20 minutes for each bottle.

    Temperature changes will depend on factors like initial water temperature, air temperature, and bottle size. Differences may be small.

15. Answers will vary. Students should notice the smallest temperature change in the bottle wrapped in two sheets of bubble wrap. It should be warmer than the other bottles after 20 minutes.

16. Answers will vary. Students may think that one or both matter. On the next page, we perform one more experiment to see whether plastic alone (using popped bubble wrap) works just as well as plastic and air (using unpopped bubble wrap).

17. Answers will vary. Students may consider Lizzie's and Alex's arguments when making their predictions. If Lizzie is right, then unpopped bubble wrap will work better than popped bubble wrap to keep the bottle warm. If Alex is right, then the popped bubble wrap should work about as well as the unpopped bubble wrap to keep the bottles warm.

18. Answers will vary. Students should record the temperature every 10 minutes using either °C or °F.

    Temperature changes will depend on factors like initial water temperature, air temperature, and bottle size. Differences may be small.

19. Answers will vary. Students should subtract the starting temperature from the final temperature to find the total change in temperature after 20 minutes for each bottle.

20. Answers may vary. We found that two layers of unpopped bubble wrap worked better than popped bubble wrap. Bubble wrap traps pockets of warm air around the bottle to slow the escape of heat.

21. The warmth in a jacket comes from **your body**.

22. The bubble wrap works by trapping warm air around the bottle, just like fur or feathers trap air to keep an animal warm.

    Air is an "insulator," meaning heat energy doesn't pass through air easily. Air that is warmed tends to stay warm longer than many other materials.

23. Similar to fur, feathers, and bubble wrap; each of these three types of insulation are made of materials that have tiny spaces filled with air that can block heat from escaping.

24. Insulation can keep something cool too. In this example, the styrofoam is blocking heat from moving from the person's hand to the snow cone. This keeps the snow cone from getting warm and melting so fast.

25. Discussions may vary. Students may argue that a cold lizard would not warm the air around itself and a sweater would not work well to keep it warm. In fact, a sweater may act like the styrofoam cup in question 24 and keep the lizard cool by blocking heat from the sun from reaching the lizard's skin.

    Students may instead argue that if the lizard were wearing a sweater after being heated by the sun all afternoon, the sweater could help the lizard stay warm into the evening as the air cools.

Chapter 16: Cooperation   292-311

1. All of the animals listed here are social and live in groups, except for koalas. Adult koalas spend most of their time alone unless they are caring for young. Although koalas do not travel around in groups, some people suggest calling more than one koala a "colony" or a "cling."

    Some groups of animals may have more than one name. We found the following group names: a troop or a barrel of monkeys, an unkindness of ravens, a gaggle of geese, a blessing of narwhals, and a swarm or a bike of hornets.

2. In the game Forage, a larger group of ants is able to carry more food. A larger group is better able to attack and defend itself from attacks.

3. In the game Forage, a larger group can not move as quickly around the board.

4. Answers will vary. Students may prefer different strategies. Students should use their experiences during the game to support their choice.

5. Answers will vary. Some students may enjoy socialization, delegation, collaboration, and other aspects of group work. Others may prefer working alone. We find that groups are most successful when each member is a useful part of the team. Groups that communicate well and support each other often have the most fun and do their best work!

6. A group of lions can capture large prey more easily than if one lion were hunting alone.

7. Lions would get more to eat if they didn't have to share their food with others.

8. When many adults care for cubs, the parents can spend more time hunting for food.

9. A large group of zebras is **more** likely to be spotted by a hungry lion. Some groups of animals are so large and easy to find, they can even be seen from space using satellites!

10. In most cases, a lion is more likely to catch a single zebra in a smaller herd where there are fewer options for the lion.

11. Wearing part of the habitat is an excellent disguise. This is an example of camouflage.

12. Answers may vary. Students should consider the differences a coral may experience moving around or being stuck in one place. As the crab searches for areas with more food, the coral may also enjoy more food. If the temperature in one place is uncomfortable, the crab may move to a more comfortable location, taking the coral with it.

13. Answers may vary. Students should consider challenges the crab may face. A crab might move more slowly if it is carrying a heavy load. The coral may face challenges if it has different needs from the crab, such as different food sources or preferred water temperature. Students might also imagine that the coral could be harmed by being removed from its original location or that the two might compete for food.

14. When mongooses stir up insects, it benefits the hornbill by making it easier to catch food. The insects do not benefit because they are eaten. The mongooses do not benefit by insects flying away.

BEETLE

MONGOOSE

HORNBILL

15. When hornbills give a warning call, it signals to the mongooses to hide and protects them from being eaten. Calling out does not benefit the hornbill, since it might attract a predator's attention. The jackal does not benefit, since its prey runs away.

MONGOOSE

HORNBILL

JACKAL

16. This sentence best describes the relationship between mongooses and hornbills.

- Hornbills are predators and mongooses are prey.
- ✓ Hornbills and mongooses help each other survive.
- Hornbills and mongooses fight over food.

17. The squirrel (A), bee (B), and boy (D) are all eating parts from plants.

18. The bee (B) helps spread pollen from one plant's flower to another.

19. The squirrel (A) is moving apple seeds to a new location. Even if it eats the apple seeds, many seeds pass safely through an animal and are left behind in its droppings.

The boy (D) is also spreading watermelon seeds. The plant that made this watermelon was likely grown on a farm far from this city park.

20. The bird (C) has built its nest in a tree to shelter its young from predators and weather.

The boy (D) is using the tree's shade to protect himself from getting too hot in the sun. Students may also note that humans build houses and other shelters from wood.

Some students might also know that squirrels (A) often live in trees and bees (B) may build their hives in trees or other plants.

21. Both the bee and the plant benefit. The bee gets some food from the plant, and the plant is able to make new seeds when the bee moves its pollen from flower to flower.

22. Discussions will vary. Most students will probably agree that an insect is more likely to get pollen to the right spot.

23. Students may discover different facts about pacific cleaner shrimp and moray eels. Moray eels open their mouths wide allowing cleaner shrimp to eat dead cells and old food from their teeth. The shrimp benefits by getting food. The eel benefits by getting cleaned.

24. Students may discover different facts about dotted humming frogs and tarantulas. These small frogs often live in or near a tarantula's nest and eat ants that would otherwise eat the tarantula's eggs. The frog benefits by getting food and also protection from the large spider. The spider benefits from the frog protecting her eggs from ants.

25. Students may discover different facts about acacia ants and acacia trees. Acacia ants make their home in acacia trees and eat its sap. If an herbivore (like a giraffe or elephant) attempts to eat the acacia's leaves, the ants will attack the animal. The ants benefit by getting food and a home from the tree. The acacia tree benefits from protection from being eaten.

Index